Venezuelan Economic Development

A Politico-Economic Analysis

CONTEMPORARY STUDIES IN ECONOMIC AND FINANCIAL ANALYSIS VOL. 7

Editors: Professor Edward I. Altman and Ingo Walter, Associate Dean
School of Business Administration, New York University

CONTEMPORARY STUDIES IN ECONOMIC AND FINANCIAL ANALYSIS

An International Series of Monographs

Series Editors: Edward I. Altman and Ingo Walter
Graduate School of Business Administration, New York University

Volume 1. DYNAMICS OF FORECASTING FINANCIAL CYCLES: THEORY, TECHNIQUE AND IMPLEMENTATION
Lacy H. Hunt, *Fidelcor, Inc. and the Fidelity Bank*

Volume 2. COMPOSITE RESERVE ASSETS IN THE INTERNATIONAL MONETARY SYSTEM
Jacob S. Dreyer, *Graduate School of Arts and Science, New York University*

Volume 3. APPLICATION OF CLASSIFICATION TECHNIQUES IN BUSINESS, BANKING AND FINANCE
Edward I. Altman, *Graduate School of Business Administration, New York University;* Robert B. Avery, *Graduate School of Industrial Administration, Carnegie-Mellon University;* Robert A. Eisenbeis and Joseph F. Sinkey, Jr., *both Federal Deposit Insurance Corporation*

Volume 4. PROBLEM AND FAILED INSTITUTIONS IN THE COMMERCIAL BANKING INDUSTRY
Joseph F. Sinkey, Jr., *Federal Deposit Insurance Corporation*

Volume 5. A NEW LOOK AT PORTFOLIO MANAGEMENT
David M. Ahlers, *Graduate School of Business and Public Administration, Cornell University*

Volume 6. MULTINATIONAL ELECTRONIC COMPANIES AND NATIONAL ECONOMIC POLICIES
Edmond Sciberras, *Science Policy Research Unit, University of Sussex*

Volume 7. VENEZUELAN ECONOMIC DEVELOPMENT: A POLITICO-ECONOMIC ANALYSIS
Loring Allen, *University of Missouri, St. Louis*

Volume 8. ENVIRONMENT, PLANNING AND THE MULTINATIONAL CORPORATION
Thomas N. Gladwin, *Centre d'Etudes Industrielles and New York University*

Volume 9. FOREIGN DIRECT INVESTMENT, INDUSTRIALIZATION AND SOCIAL CHANGE
Stephen J. Kobrin, *Alfred P. Sloan School of Management, Massachusetts Institute of Technology*

Volume 10. IMPLICATIONS OF REGULATION ON BANK EXPANSION: A SIMULATION ANALYSIS
George Oldfield, Jr., *The Amos Tuck School of Business Adminstration, Dartmouth College*

Volume 11. IMPORT SUBSTITUTION, TRADE AND DEVELOPMENT
Jaleel Ahmad, *Concordia University, Sir George Williams Campus*

Volume 12. CORPORATE GROWTH AND COMMON STOCK RISK
David Fewings, *McGill University*

Volume 13. CAPITAL MARKET EQUILIBRIUM AND CORPORATE FINANCIAL DECISIONS
Richard C. Stapleton, *Manchester Business School, England and New York University and* M.G. Subrahmanyam, *Indian Institute of Management, Ahmedbad and New York University*

Volume 14. IMPENDING CHANGES FOR SECURITIES MARKETS: WHAT ROLE FOR THE EXCHANGES?
Ernest Bloch and Robert A. Schwartz, *Graduate School of Business Administration, New York University*

Volume 15. INDUSTRIAL INNOVATION AND INTERNATIONAL TRADING PERFORMANCE
William B. Walker, *Science Policy Research Unit, University of Sussex*

Volume 16. FINANCIAL POLICY, INFLATION AND ECONOMIC DEVELOPMENT: THE MEXICAN EXPERIENCE
John K. Thompson, *New York University*

Volume 17. THE MANAGEMENT OF CORPORATE LIQUIDITY
Michel Levasseur, *Centre d'Enseignement Superieur des Affaires, Jouy-en-Josas, France*

Volume 18. STABILIZATION OF INTERNATIONAL COMMODITY MARKETS
Paul Hallwood, *University of Aberdeen*

Volume 19. PROBLEMS IN EUROPEAN CORPORATE FINANCE: TEXT AND CASES
Michel Schlosser, *Centre d'Enseignement Superieur des Affaires, Jouy-en-Josas, France*

Volume 20. THE MULTINATIONAL ENTERPRISE: INTERNATIONAL INVESTMENT AND HOST COUNTRY IMPACTS
Thomas G. Parry, *School of Economics, University of New South Wales*

Venezuelan Economic Development
A Politico-Economic Analysis

by LORING ALLEN
Department of Economics
University of Missouri, St. Louis

JAI PRESS
Greenwich, Connecticut

Library of Congress Cataloging in Publication Data

Allen, Loring.
Venezuelan economic development.

(Contemporary studies in economic and financial analysis ; v. 7)
Bibliography: p.
Includes index.
1. Venezuela—Economic conditions. 2. Venezuela—Politics and government. 3. Petroleum industry and trade—Venezuela. I. Title. II. Series.
HC237.A65 330.9'87'063 76-10395
ISBN 0-89232-011-7

321 Greenwich Avenue
Greenwich, Connecticut 06830

ISBN: 0-89232-011-7
Library of Congress Catalog Card Number 76-10395
Manufactured in the United States of America

To
Jeanne Miriam,
Robert Loring II
and
Maris Jean

ACKNOWLEDGMENTS

My first professional activity in Venezuela was in 1962 and from that time until 1968 I worked with the Universidad de Los Andes in Mérida as a visiting professor sponsored by the University of Oregon and the Ford Foundation. Subsequently, the Research Office and the Center for International Studies of the University of Missouri-St. Louis assisted in my annual trips to Venezuela. I was also a Fulbright lecturer at the Universidad Simón Bolívar in Caracas in 1975.

My debts to Venezuela and the United States are legion. To the Facultad de Economía of the Universidad de Los Andes I feel a special bond. To its professors and students I owe my interest in the Venezuelan economy and its development. One name stands above the rest: Dr. Heberto Urdaneta, former dean and now general manager of the Venezuelan Investment Fund, who has guided and helped me in many ways. But an equal obligation is due to many others who in classroom and cafetín struggled to enlighten a benighted gringo professor.

Several Venezuelan colleagues read earlier versions of this book and suggested many improvements. These include Dr. Bernardo Ferrán, Banco Central de Venezuela; Dr. Gorgias Garriga, Caracas Petroleum Company; Drs. Enrique Tejera Paris and Antonio Blavia, Universidad Simón Bolívar. Among North Americans, Professors Robert Alexander of Rutgers; Winfield Burggraaff of the University of Missouri-Columbia; and Hugh Nourse of the University of Missouri-St. Louis, were most helpful.

One of the greatest pleasures in writing the book was working with these and many other people, both North American and Venezuelan. But I will not be surprised if some of these friends disavow some of the interpretations, and I would be remiss if I did not excuse them from all culpability.

LORING ALLEN

CONTENTS

III Evolution of Democracy

IV Petroleum Policy

V Economic Policies

LIST OF TABLES

Venezuelan Economic Development

A Politico-Economic Analysis

Chapter I

The Venezuelan Setting

The Venezuelan nation stands on the summit of success in 1975. On every side evidence of maturity, prosperity and promise brightens the economic and political landscape. Petroleum policy has finally brought that behemoth into the public domain through nationalization after decades of mounting public participation. The petroleum price increases of 1973 and 1974, for which Venezuela may claim some credit, have provided the financial resources for economic growth and diversification. Economic and financial policies have yielded the highest growth rate and per capita income in Latin America and a sound basis for expansion. Democratic processes function smoothly yet vigorously.

On the other hand, Venezuela was a pathetic country in 1935. The dictator, General Juan Vicente Gómez, had ruled with absolute authority for twenty-seven years. Only the favorites of the dictator prospered. The small petroleum enclave benefitted foreigners and the dictator. Most Venezuelans eked out a mean living from the soil. The cities were dreary, the countryside bleak, the people disheartened. The twentieth century had not arrived in Venezuela.

Between 1935 and 1975 change convulsed this country. Not the ordinary change which often leaves many things the same, but a searing, rending economic and political upheaval in which almost nothing remained the same. It was a metamorphosis, a transformation in which the form and content of the Venezuela of 1975 resembled only remotely the Venezuela of 1935. Some might call it a miracle or a revolution; elements of both were in the burst of energy which created today's Venezuela.

In 1935 the economy was dependent on petroleum. In 1975 the economy was still dependent on petroleum. The nature of the dependence, however, had changed. Recognized and government-managed dependence had replaced the earlier uncontrolled dependence. Public policies eliminated some kinds of dependence, including private, foreign decision-making in the petroleum sector. Other kinds of dependence, such as trade, fiscal and technological dependence remained, but under more control. The transformation of the last forty years has moved Venezuela closer to interdependence.

The transformation has two principal dimensions: economic and political. A vegetating, malformed and malfunctioning economy became a sturdy and growing economy. A ruthless dictatorship became a workable democracy. Petroleum was the engine of change; petroleum policies converted the oil into revenues that economic and financial policies put to work to propel the economy into modernity. All of these components interacted to produce the present politico-economic system of Venezuela. This book analyzes the interaction of petroleum, politics, policies and performance in the context of changing dependence in the last generation.[1]

The Saga of Juan Bimba

Consider the case of Juan Bimba, the national caricature: this lean and hungry-looking Venezuelan was born in the year the dictator died. In the early years he lived in a shack on the side of a hill with his father and mother and six other children. As soon as he was able, he helped his father farm a little plot which provided the family with little more than subsistence. He did not go to school because he had to help his father and anyway, the nearest school was eighteen kilometers away. As a teenager his social life consisted of taking produce to the small city on Saturday and getting drunk on the proceeds. On one such excursion he met a girl and she became pregnant. Thus, in 1950, the odyssey of Juan Bimba begins.[2]

He moved to town and married the girl. It was the first and last time he was ever in church; yet he was a devout Catholic. He got a job driving a por puesto, a car that runs a regular route through the city streets. He lived in a little shack under the bridge on the outskirts of town. The years rolled by and each year he earned a bit more and his family grew larger. His ranchito also grew; first one more room, then another, made space for the annual new child. Pictures appeared on the wall; he bought a radio. Juan worked long hours and he spent his spare time repairing the car. The year 1958 was a red letter year. He bought, on credit, a half-interest in a car, his eldest son started going to school, and the family moved to a new barrio where his ranchito had a wooden floor.

In the country, all attention centered on Caracas. A new temporary military junta had ousted the dictator and promised free elections. Juan and his partner and their families moved to Caracas, in the hope of earning higher pay. The first years were hard. The country's economy was in depression and Juan's income fell instead of rising. His housing was worse, many of his friends could not find jobs, and Juan had a difficult time keeping his children in school. But then things began to get a little better.

Juan became politicized. He worked hard in his spare time for Acción Democrática. In 1963 his big break came. Through political friends he got a job driving a truck for the Ministry of Public Works at twice the pay he had been receiving. He sold his interest in the por puesto. Now he had his evenings free. At home he felt pressure from his children who could read and write. He went to night school, learned to read a newspaper and to write a few words. Deciphering the words and drawings of his truck technical manual was an arduous chore but he accomplished it. He avidly supported his party and attended all the big meetings. He managed to keep his children in school through the sixth grade. One, the eldest, was even going on.

Juan's political and economic sentiments were primitive and not well informed. He thought poor people, like himself, should have more. He couldn't understand why the United States sent its big crooked companies to his country and took all the oil without giving something in return. Maybe that's how the Yankees got so rich, he reasoned, taking things from other people. But anyway, he knew how to get along: do what you're told, know the right people, and be lucky. He believed in his party but he knew the government was corrupt. Those not being paid off by the Yankees were bought by Mendoza or some other rich man. Sometimes he daydreamed about being given a lot of money just for knowing the right people.

Again through political friends an opportunity arose. He became a timekeeper on a big project. He talked to engineers and plumbers and electricians. He began to notice that the people who were the most able and worked the hardest made more money and got promotions. With time on his hands even at work he started reading the newspaper and some school books. He began to shave every day and wore better clothes to work. The idea began to grow in him that he was able to supervise a group of workers, that by knowing several jobs and working very hard, he could earn more pay. He studied what other people did. He moved out of the barrio into an apartment with water and a toilet. In 1970 his son graduated from liceo and entered the university to study engineering.

Finally, his efforts attracted the attention of the engineer in charge. He promoted Juan to straw boss, responsible for six men on a cement crew. Juan's interest in politics dwindled but he remained a good party man. He was happy that his party was going to throw the North American companies out and take over the oil. If Venezuelans could build buildings and roads and a subway, they could run the petroleum industry, he thought. He fired two men from his crew; nothing unusual in that, except that they were from his own political party. It made him uncomfortable to do it, but the men were just not working hard enough. In 1975 his son graduated from the university; Juan was in charge of three crews. For the first time he had to pay income taxes. Juan Bimba had made it.

From Colonial Town to Modern Metropolis

Consider also the case of a small town up in the Andes Mountains. In 1935 it was a charming but very poor little town, isolated from the rest of the country, with only a corkscrew mountain road connecting it to nearby cities. The prisoners of Gómez had built the road. There was electricity, but the candlemaker and kerosene vendor did a brisk business. The town was small, all huddled around Plaza Bolívar, its gargoyled cathedral, and a clutch of shops. When the dictator died, the town's intellectuals began to think about reopening the university.

The town only had a few streets and few were paved. Only the very rich had cars, and the occasional out-of-town car was an event. Only a few children went to school because so few schools existed. The rich doctors and merchants sent their children to school in Caracas. The houses were all colonial style, with open patios in the middle. Going up and down the street one could hardly detect that this was well into the twentieth century, not the eighteenth.

Bit by bit the changes came. A new road, straighter and less dangerous, was driven through the mountains and automobile traffic increased. The government laid out an airport and regular flights connected the town to Caracas and other cities. Electric signs began to appear outside stores. Even neon lights. Imported goods, displayed behind plateglass storefronts, became ubiquitous. The town built more roads and paved some of the old ones. Apartment houses blossomed and many new-style houses appeared. Agriculture thrived and enterprising merchants constructed small factories. A fancy modern hotel with a swimming pool drew tourists from Caracas.

Population expanded rapidly as people from smaller towns and from the countryside arrived. More and more children were in the streets. New as well as antiquated school buildings burst at the seams with scrubbed, white-smocked children. The local university came to life, expanded and introduced new studies. Even foreigners, babbling strange languages, began teaching at the university. The city built a park honoring its poets and writers. More houses, streets, stores, goods and people appeared.

Today it is worth your life to venture across the street. An equestrian Bolívar looks down on traffic jams. Blaring horns and hawking newsboys assault the ear. Stoplights feebly regulate traffic. Neon lights announce laundromats, tobacco stores and restaurants. Jazz music comes from shirt pockets, television antennae rise even from the ranchitos, and parking meters proclaim the arrival of civilization. In a generation the sleepy little colonial town has become a bustling and overcrowded modern city.

The same thing happened all across Venezuela. Caracas was home to 200,000 people in 1935. Nearly two and one-half million lived there in 1975 and their incomes had skyrocketed, despite the ups and downs and the inequality. In 1941 nearly 70 percent of the people lived in the countryside; by 1975 less than 25 percent were still there. Out in the Guayana two little villages numbering less than 4,000 nestled in the juncture of the Orinoco and Caroní Rivers in 1950. In 1975 they had become a new city of nearly 150,000, the center of a huge industrial complex.

Growth and change are manifest everywhere, in the stores and streets, in the houses and people, and in what the people do. The poor, the ill-fed, ill-housed and ill-clothed are still there, in even larger numbers, but less obvious in the frantic economic activity all around them. Oil-borne prosperity and economic growth has conferred on Venezuelans a livelihood heretofore unknown in Latin America, but still has not resolved the basic economic problems of poverty and underdevelopment. Despite its high income, Venezuela is still only a partly developed country, heavily dependent on purchases by industrial countries of a single product, and still divided inside the country between the haves and have-nots.

Physical Characteristics

Venezuela is a land of jungle darkness and bright city lights, full of contradictions and paradoxes. Its per capita production in 1975 was nearly $2,500, much higher than that of any country in the Western Hemisphere save for the United States and Canada. Yet the ranchitos—the miserable huts outside Caracas and other cities—bespeak a poverty so abysmal that it must be seen to be believed. The per capita income of the poorer states is no more than one-tenth of the national average. Some Indian villagers live in pre-Colombian backwardness. In other areas life has changed little since the seventeenth century. Yet the glamorous and glittering cities with shiny cars, modern freeways, and sophisticated restaurants as well as the gay night life are envied in Europe and the United States.

Venezuela forms an irregular triangle with its apex standing almost on the equator and the base line running along the Caribbean Sea. It occupies the northeast shoulder of South America, south of the eastern tip of the United States. North to south it measures more than 900 miles and east to west 800 miles, bordering on Colombia, Brazil and Guyana, with 1,750 miles of coast along the Atlantic and Caribbean. It is the seventh largest country in Latin America, about 10 percent of the area of the United States.

A part of the national territory is the Guayana shield, one of the earth's oldest physical features, 2 billion years old. As late as 90 million years ago the sea covered much of Venezuela, that portion north and west of the shield. The present land mass rose from the sea to form the Andes and Coastal mountain ranges, and then the great llanos, or plains, filled in. In the last million years the llanos have continued to rise.[3] Of the four natural regions, the largest—45 percent of the nation—is the remnant of the Guayana shield. It plays only a small part in Venezuelan economic life and has less than one percent of the population. High plateaus, deep gorges cut by swift streams, and scattered savannas dominate the region. Its dark forests sheltered Abel and Rima in Hudson's classic *Green Mansions*.[4] Jimmy Angel discovered the highest waterfalls in the world in the Guayana in 1934. Most of the territory remains unexplored. An industrial complex in the northeastern part of the region, along the Caroní River, is the chief connection with the modern economy.

The llanos occupy more than one-half of the remaining national territory. They lie north of the Orinoco River, extending from inside Colombia on the west to the northeast coast of Venezuela, 200 miles wide and 1,000 miles long. An inland sea in the wet season, the llanos are parched and insufferably hot during the dry season. Cattle, often wild, outnumber people ten to one. The contribution of the region to national economic life is limited.

Dividing the llanos and the Guayana is the Orinoco River. With its tributaries it drains the northern and western mountains beyond the llanos as well as the Guayana, more than 80 percent of the nation. Slow moving in its tortuous path, it runs 1,300 miles, the second longest river in South America. Winding through empty land it has little economic significance, although along its shores are many small settlements. Near its delta are two major population centers: Ciudad Bolívar, and at the juncture of the Orinoco and Caroní, Ciudad Guayana (Santo Tomé de la

Guayana), the industrial center. The delta of the Orinoco fans out and the river drains through many mouths into the Atlantic.

To the northwest, on the other side of the Andes, is another lowland area, the Maracaibo basin, dominated by Lake Maracaibo, the tear-drop shaped lake which is the largest in South America. It connects with the Caribbean through the Gulf of Venezuela and drains much of the seaside northern Andes. The lake nestles between two trunks of these mountains. Lake Maracaibo did not enter the economic life of the country, except as transportation, until the twentieth century with the discovery of petroleum under and around the lake. The basin, a fertile agricultural region, comprises seven percent of the nation.

The mountain zone, extending from the Colombian border in the southwest to the Peninsula of Paria in the northeast, is the most important economic region. It is also of great historical importance, the principal settled zone in colonial days; the birthplace of Independence; and the agricultural, industrial and population heartland of the nation. Caracas is in a mountain valley along the northern coast.

The social, political and cultural fabric of the nation has always centered in the mountain valleys of the Andes, the north central, and the northwest. Two mountain ranges, almost touching, are the spinal column of Venezuela. One is the right spur of the northern Andes chain. It enters Venezuela from Colombia, where one branch goes almost directly north to the sea, west of Lake Maracaibo. The Cordillera de Mérida, after a low saddle near the border, moves northeast, to the south and east of the lake.

The Andes vary in width from 10 to 40 miles and include many small valleys and high mountain grasslands. These are important in Venezuelan agriculture. A number of prominent peaks, notably Pico Bolívar (16,427 feet), the highest in Venezuela, overlook the towns, roads and agricultural areas below. The Coastal range begins a short distance from the end of the Andes and runs parallel to the coast. It consists of two sierras, separated by a valley up to 30 miles wide, another significant agricultural zone.

Venezuela lies wholly in the tropical zone, and at sea level the heat is unbearable. Temperature variation is a function of elevation and the mountain valleys are comfortable. Freezing conditions prevail only in the high Andes. May to November is winter, because of the heavier precipitation, while the rest of the year is warmer and drier. Humidity runs high, between 65 and 80 percent. Rainfall ranges from 20 to 80 inches, heaviest in the mountains and the Orinoco delta.

Only three percent of the land is arable, mainly the alluvial soils of the mountain valleys. About 20 percent is under forests and another 20 percent is in pasture. Much of the land has not been used and may not be usable, consisting of swamps, inland water, high mountain areas, scrub forest, mangrove thickets, and semi-desert.[5] Slash and burn agriculture and seasonal rains over the centuries have caused great erosion and loss of arable land. Population centers have long since driven wild life to the south where tropical birds, cats and other animals still abound.

Social Characteristics

In a grueling, century-long conflict the Spaniards took the territory of Venezuela from a dozen scattered Indian tribes. The Spainard mixed with the Indian, and later the black slave added to the miscegenation. The mestizo population today dominates all aspects of the country's economic, political, social and cultural life. The social and cultural structure continues to reflect the Hispanic tradition, but with increasing Anglo-Saxon influence.

The upper socio-economic class has two elements: an oligarchy of land-owners whose wealth comes from the production of coffee, cacao and other crops, and a newer oligarchy whose riches originate in commerce and industry. Characteristically, the upper class has traditionally had the Spanish disdain for work, particularly manual effort. The infusion from the industrial oligarchy has begun to change this. The middle class, still small but multiplying rapidly, began with the oil boom in the 1920s. It consists of small businessmen, professionals, managerial and technical people, and the military and governmental bureaucracy. Since status in the middle class has usually resulted from effort, these people tend to be more achievement-oriented.

The lower socio-economic class (numerically far superior) has little group sentiment other than the rather vague hope of economic improvement. Since the middle class recruits from the lower class of previous generations, most people low on the socio-economic ladder live in the expectation that they, or their offspring, will also experience significant economic betterment.

The demonstration effect, through the media, personal conduct of leaders, and political promises, is strong and constant pressure bears down on the political system to improve conditions of low income groups.[6] A dual wage structure prevails. Wages in the petroleum industry, emulated in part by wages in industry, are high and sufficient to place these workers in the middle class. On the other hand, in agriculture and many service and commercial activities wages are low. Unemployment and underemployment are both serious problems, in the city and countryside.[7]

One of the most grave social problems is the housing deficit, estimated in the 1970s at 800,000 dwelling units. Were the substandard ranchitos torn down, the deficit would be much greater. The shortage arises from the massive migration to the cities, the high rate of population growth, and modest facilities for housing credit. The government pours substantial resources into housing each year but the deficit remains.

On the other hand, great progress is evident in education at all levels. Education did not fare well under the dictatorships, but beginning in 1959 democratic governments have given the highest priority to education, and in the 1970s nearly one-fourth of the population goes to school. Venezuela devotes a higher proportion of total product to education than does any country in Latin America. Education is free and compulsory to age fourteen, and even university education is free. Special programs have made deep inroads into the earlier widespread illiteracy and adult education offers drop-outs the chance to complete their education.

University education is becoming more technically oriented, with schools of science, engineering, economics and business administration prospering. In 1974 President Pérez pledged an advanced education abroad to 10,000 Venezuelans a year, with emphasis on specialized technical training. Despite these efforts a shortage of classrooms and qualified teachers prevails, drop-out rates are high, and the quality of education is often inferior.

The major socio-economic problem is deep and entrenched. It is the unequal distribution of land, wealth and income from which is derived the unequal distribution of economic and political power. The rapid growth of the last forty years has tended to improve the conditions of all classes, but has not improved much the distribution of money income. Since 1959 the government has introduced many measures to ameliorate the inequality and has effected some improvement. The allocation of the benefits of growth among socio-economic groups remains, however, the critical issue.

Venezuela comes from a social and intellectual tradition quite different from that which produced the phenomenal economic progress of the last two centuries. The country's colonial and national development was not supportive of a sense of pragmatism, social and economic egalitarianism, scientific inquiry, or of a reward system based on individual capability and performance. The Venezuelan historical experience indeed was counterproductive of economic development. Since 1935 this new ethos has begun to emerge, and achievement-oriented behavior and a developmental outlook have gained common currency.

Common Growth Patterns

Scholars are forever trying to make the particular fit the general. The growth pattern of one country at one time must be made to fit into a common pattern for all countries at all times. It is a part of the eternal search for the general rule. Venezuela has not escaped the attention of those who wish to apply general explanations and development patterns. Their efforts have been rewarded by a number of stereotypes concerning the economy which many inside and outside Venezuela accept.

Venezuela has its share of Marxist analysts who interpret Venezuelan economic development according to that dogma. Most of the analysis strains the spirit and letter of Marx since he dealt with the central core of capitalist industrial development, not with peripheral monocultural areas. Marx did, however, deal with colonies of capitalist countries, and Venezuela, as a Spanish colony and more recently as a supplier of a primary product, is a colony according to latter-day Marxists. Indeed, in many respects, until the emergence of more nationalistic economic policies, Venezuela seemed to fit the exploitation and dependency model of Marx. Many today consider Venezuela at least a neo-colonial economy, controlled by industrial countries that extract the country's wealth with inadequate payment and prevent industrialization with cheap industrial exports.[8]

The Venezuelan domestic economy also operates according to the model of Marx, some argue. A relatively small number of families and enterprises control the wealth of the country and the economy. They are not quite the industrial capitalists of Marx, but they are great concentrations of power that dominate industry, construction, banking, much of commerce, and they influence the government as well. While much of the Marxist analysis does not conform well to many aspects of the Venezuelan reality, a reasonable facsimile can be put together with the generous use of Marxist jargon as a substitute for analysis.[9]

The Marxist prognosis, however, is not appropriate. The course of events and the fate of Venezuela depend only in part upon what happens in Venezuela. Rather, it depends on the development of industrial countries. Whatever course these countries may be on, they do not seem to fulfill the Marxist prophecy of collapse and socialism. Increasing governmental interference with capitalist processes and dominant economic policies may introduce a form of socialism without the necessity of capitalist collapse and violent introduction of socialism.

Developments within the Venezuelan domestic economy indicate that it is following a path of government control and planning, egalitarian policies, and diminished influence by private decision-making. Whatever appeal Marxist ideology may have, Marxist economic analysis of the development process contributes little enlightenment to what has been happening in Venezuela in the past forty years.

One of the most popular recent studies of economic growth is that of Walt W. Rostow.[10] It has aroused much discussion, controversy and speculation. A favorite exercise of some scholars is to try to fit some country at some time into one or another of Rostow's five stages. In his traditional stage, productivity is unchanging and there is no systematic economic progress. In the pre-condition stage, the forces of change gather strength, some progress occurs, and the economy lays the foundation for the third and critical stage—the take-off, when the economy explodes out of its past and achieves a new and higher rate of sustained economic growth. In the fifth stage the economy achieves prosperity and high mass consumption. This pattern of growth fits reasonably well into the English development, some continental experiences, and the North American case.

A number of observers have argued that Venezuela entered the vital take-off stage in the late 1950s or early 1960s. Alfredo Machado Gómez, in his interpretation of the Venezuelan economy says that Venezuela met the take-off requirements in the 1950s, and that in 1960 the country neared the stage of sustained growth.[11] John Friedman states that "the situation during the fifties and sixties appears to fit Rostow's criteria for the 'take-off' reasonably well."[12]

Even Rostow himself argued that the take-off was in process in Venezuela in the 1960s. He said, in 1962 that, "in the past few years Venezuela may have moved over into the category of economies experiencing an authentic take-off."[13] In addition, the Economic Commission for Latin America asserts that "the decade of the 1940s may be judged by future historians as the take-off period of the Venezuelan economy."[14]

The facts, however, do not fully support the take-off hypothesis for Venezuela in the last generation. The rapid growth before 1958 and in the 1970s originated in external conditions, just as did the slower growth of the 1960s. Domestic economic policies heightened the growth impact of external factors but they neither created nor destroyed the externally-conditioned growth. The 1960s reflected nearly self-sustained growth but petroleum still supported the economy, just as it generated the accelerated growth of the 1970s.

The central features of the last forty years in the economy were growth and structural transformation. It was not independent growth, nor self-sustaining, nor regular. Growth was acutely irregular, and faltered on more than one occasion. After the orgy of investment of the 1950s (much of it contributing little to the productive forces of the economy) investment declined in the 1960s, making, however, a relatively larger contribution to the accomplishment of the government's goals. The leading sector of the 1950s—petroleum—fell behind in the 1960s, but industry did not capture the lead; services was the leader. And petroleum once again became the leading sector in the 1970s, as a result of price increases.

Venezuela may or may not have taken-off in the 1950s and 1960s, or even in the 1970s. The question is whether or not the stages of growth schemata, including the take-off, is the appropriate framework for the analysis of the Venezuelan economy. The course of economic and political events, economic growth, political and social conditions, sectoral relationships, contacts with the exterior, the institutional framework, and public policy suggest that Venezuela was in all the stages of growth simultaneously. Alternatively, if one wishes to call the transformation of the last forty years the take-off, then it was its own special kind of take-off, unlike any others in its origin, course and consequences.

The paradox of external dependence and influence of domestic policy calls into question the applicability of the take-off phenomenon in the Venezuelan economy. What happened was in part beyond the control of Venezuela, and in part the consequence of policies undertaken to reduce dependence and fortify long-run growth prospects. No country whose principal revenues derive from a single product that contributes one-fourth or more of total production can take-off independently. Policies may, and in the Venezuelan case have, changed and managed the dependence, but have not eliminated it. Venezuela grew rapidly or slowly, smoothly or irregularly, depending on forces beyond its control.

At the same time, the government exercised some influence on the rate of growth. Venezuela grew rapidly up to 1958 not only because of favorable external circumstances but also because the government played successfully upon these external factors in order to grow rapidly. In doing so, dependence deepened. In the 1960s, Venezuela pursued policies which, in view of changed external conditions, produced even slower growth than that which would have resulted from external factors. In employing these policies, dependence changed; the country deliberately sacrificed growth for greater independence. In the 1970s the huge revenues produced by the policies of independence, in conjunction with petroleum price increases, had a decisive impact on growth, but perpetuated the dependence of the economy.

Douglas North dealt with the critical role of an export staple in inducing capital formation and the development of secondary and tertiary sectors, as well as other export staples. Capital formation and expansion of the export staple generates increasing income, permits saving and a higher rate of investment in the non-staple sectors, broadens the market, and thus produces economic diversification and development. Under some circumstances this may lead to industrialization, in others to the development of new export staples. North was writing in the context of North American regional development, but even earlier Harold Innes had endeavored to explain Canadian economic history in these terms.[15]

Certainly Venezuela has a staple—petroleum—and its role has been decisive in its history since 1920. The Venezuelan case was, however, rendered unique by the fact that foreign factors of production developed the petroleum sector in the context of limited linkages to the domestic economy. Petroleum developed and operated as a foreign enclave, touching the domestic economy at only a few points. In 1950 petroleum employed only three percent of the labor force, in 1970 less than one percent. The capital came exclusively from abroad and in 1975, before nationalization, no significant domestic petroleum industry existed. Indeed, petroleum was not a sector of the Venezuelan economy. It was a foreign sector in Venezuela with a highly specific stimulating effect on the domestic economy.

The principal contact of petroleum with the domestic economy was through the government. Taxes were small in the early years, however, and not until the late 1940s did the government share even equally in company profits. Through most of the 1950s the shares were about equal and only since 1958 has the government received more than the foreign owners.

In addition, only a fraction of the government share went to capital formation and even then primarily for infrastructure, not for productive activities. The petroleum-government nexus generated few incomes sufficiently large to produce enough savings to influence the domestic rate of investment so that independent development took place. While the export staple theory does not fit the Venezuelan case well in the past, it is useful in emphasizing the differences between domestic and foreign export staple development and the domestic economy and may be more appropriate in the 1970s when, through nationalization, petroleum may have an increasingly significant domestic impact.

Colin Clark traced economic development through the evolution of the sectors of the economy. After an initial expansion of the primary sector, mainly agriculture, as reflected in share of output and use of labor, a country industrializes on the basis of profits from primary production. The secondary sector provides, for a period, a growing share of total output and employment. Subsequently, the share in the secondary sector stabilizes or declines and the tertiary share increases.[16] This, indeed, depicts the manner in which most present-day industrial countries have developed.

This orderly sector progression has not transpired in Venezuela, or, in fact, in most less developed countries. In Venezuela the primary sector share has declined in real terms, but it was petroleum which declined. Industrial production expanded its share modestly; there was, however, no great industrialization drive. The tertiary sector expanded rapidly and the expansion of its share matched the decline in the petroleum share. The profits of Venezuela's primary sector went abroad, or to the government which promoted tertiary development. With the petroleum price increase, the primary share has now increased sharply again, in money but not real terms.

The Venezuelan case does not fit any of the global concepts of economic development precisely, but conforms to all of them in part. The transformation of the Venezuelan economy cannot be explained by the historical experience of already developed countries. Nearly all of these countries in their process of development, managed to extricate themselves from economic dependence, to convert dependence into interdependence. Despite the changes in the Venezuelan economy, dependence, though altered in form and under more control, still prevails.

Economic Dependence

Before the discovery of Venezuela, the diffuse Indian civilization had maintained a diversified if primitive economy, making progress and capable of feeding itself. Since the conquest pearls, slaves, tobacco, hides, cacao, coffee and finally petroleum have in succession held the economy in their grip. While the economy has progressed, especially in the last forty years, economic dependence has distorted production patterns and the distribution of income, creating and perpetuating economic underdevelopment. Only since 1958, with a new political regime, has the country made efforts to extricate itself from dependence and to manage that dependence which remains.

Deeply ingrained in the thinking of many Venezuelans is the belief that they are impotent in the face of forces beyond their control, that they are not and have never been really in charge of their own affairs. They point to the oil companies, the German coffee merchants, the effects of restrictions of oil-buying countries in the 1960s, the ubiquitous foreign goods, United States and foreign investment, huge payments for foreign technology, and many other indicators of dependence. This feeling of dependence underlies much of the public policy of the country.

Analysts in less developed countries have spun elaborate theories about the notion of economic dependence while those of the industrialized countries view the concept with suspicion. The former regard it as a part of the unconscious exploitation practiced by the strong on the weak. The latter point to its lack of precision, the interdependence of all trading nations, and the advantages of trade and foreign investment.[17]

It is not sufficient to dismiss economic dependence with the assertion that markets are interdependent, that buyers depend on sellers and sellers on buyers. American consumers buy fuel oil from Venezuelan producers; they are mutually dependent. Not so. This ignores who the actors are in international markets, the imperfect market apparatus, how transactions actually take place, and the use and abuse of economic and political power.

American consumers do not buy Venezuelan oil. Large companies sell oil to consumers and buy the oil from their own subsidiaries in Venezuela; in effect, American, British and Dutch producers sell to themselves and to enterprises they control. A few giant, interdependent enterprises control the extraction, production, processing and marketing of physical assets of Venezuela and nearly all other nations in which oil deposits exist. The economic power of the Venezuelan nation does not rival that of some of the companies which operated there before nationalization.

Nor is it appropriate to argue that those who employ the notion of economic dependence are simply seeking a scapegoat, a way to explain their own failure to wrest from niggardly nature its bounties. The North American colonies depended economically on Great Britain but they, by their own efforts, it is asserted, developed a great and rich nation. And they had no oil to start them on the way. So why does not Venezuela with its fabulous mineral wealth work hard, save, be responsible, and become a great and rich nation? Such analogies have limited applicability and ignore complex historical interrelationships.

It is difficult to define economic dependence precisely. It is a situation in which decision-making with respect to important aspects of the economy is not in the hands of its nationals nor is it subject to significant influence by the country's government. Armando Córdoba defines it as "the subordination of one economy which has established a system of relationships of such a nature that institutionalizes the subordination and impedes the possibility of the exercise of similar actions in the opposite direction."[18]

Economic dependence is a specific but varied phenomenon which operates at the level of particular economic relationships. Venezuela is dependent on its petroleum sector. Until nationalization, this dependence had at least eight dimensions, including export and import dependence, revenue dependence, enterprise decision-making dependence, technological dependence, dependence reflecting the great size of the industry relative to the economy, and dependence originating from past policies of the enterprises. Each of these types could be subdivided. Nationalization will eliminate some kinds of dependence, but will leave others untouched. Since some of the dependence originates from the specialization of the Venezuelan economy, it would not be possible or even desirable to eliminate it.

Economic dependence does not necessarily have anything to do with the division of benefits from trade, investment, or other economic relations. An econony may receive substantial economic benefits from the dependency, the benefits of specialization, which simulataneously leave the economy vulnerable. It is possible that the economic benefits can exceed those in which the dependence does not exist. Historically, it must be admitted, that the benefits to those who are dependent have been less than they would have been in the absence of dependence. Those with power tend to use that power to enrich themselves and gain more power.

If the division of benefits is not the central issue, then what difference does it make? The difference is decision-making itself. Decision-making has economic and political value and those who do not have it are less well off than they would have been with the ability to make decisions. Economic dependence deprives the dependent of a benefit enjoyed by others. While the comparison is extreme, economic dependence shares some of the nature of slavery.

The concept lacks precision. What is economic dependence at one time, place, or situation may not be in another. It is also a question of degree. A shift in some underlying condition may convert dependence into normal economic relations and vice versa. Dependence is also subject to manipulation and management by public policy. Nevertheless, dependence is real. Many nations, including Venezuela, are caught up in dependent relations far beyond those implied by normal trade and investment. This does not mean that a giant conspiracy among enterprises exists, that markets are rigged, or that the powerful deliberately foist dependence and exploitation on the weak, although all these have happened. Economic dependence is simply a reflection of the fact that power relationships are not equal.

The first centuries of the Venezuelan experience illustrate its dependence, depict how the politico-economic system evolved into the twentieth century, and set the stage for the analysis of the last generation. The next chapter treats briefly the first twelve generations of Venezuela, from pearls to petroleum.

NOTES

1. The most important source of statistical data concerning every aspect of the Venezuelan economy is the Banco Central de Venezuela. In its annual *Memorias* and *Informes Económicos* it publishes data compiled not only by its own department of economic studies but also data from ministries and other governmental units. It maintains the official income and product accounts of the nation. The Dirección General de Estadística y Censos Nacionales publishes all census reports as well as the annual *Anuario Estadístico* and other special reports. Each ministry and governmental unit publishes an annual report, including a statistical annex which covers the sector of the economy under its jurisdiction. One of the most important is *Petróleo y Otros Datos Estadísticos*, published by the Ministry of Mines and Hydrocarbons.

A number of recent books in English deal with special aspects of Venezuelan economic and political development. These include David Eugene Blank, *Politics in Venezuela*, 1973; Winfield J. Burggraaff, *The Venezuelan Armed Forces, 1935–1959*, 1972; Mostafa F. Hassan, *Economic Growth and Employment Problems in Venezuela*, 1975; David H. Levine, *Confiict and Political Change in Venezuela*, 1973; John Duncan Powell, *Political Mobilization of the Venezuelan Peasant*, 1971; Franklin Tugwell, *The Politics of Oil in Venezuela*, 1975; Luis Vallenilla, *Oil: The Making of a New Economic Order*, 1975; and Thomas A. Weil and others, *Area Handbook for Venezuela*, 1971. The complete citation for these works and other sources is given in the Bibliography.

2. Raymond Crist and Edward Leahy. *Venezuela: Search for a Middle Ground*, 1969, contains four illuminating profiles of typical Venezuelans in various walks of life.

3. Pablo Vila. *Geografía de Venezuela*, 1969, and his *Visiones Geohistŕicas de Venezuela*, 1969, are authoritative studies of Venezuelan geography.

4. The Venezuela Guayana was the setting of William Henry Hudson's famous masterpiece, one of the jewels of English prose, *Green Mansions*, New York: Grosset and Dunlap, 1916, Universal Library edition.

5. Luis Fernando Chaves. *Geografía Agraria de Venezuela*, 1963, Chapter III; Thomas J. Abercrombie. "Venezuela Builds on Oil," *National Geographic*, March 1963, is a colorful panorama of the Venezuelan scene.

6. Lisa Redfield Peattie. *The View from the Barrio*, 1968, is a detailed description of barrio daily life by someone who lived there.

7. Chi-Yi Chen. *Economía Social de Trabajo*, 1969, Chapter VI; and Carlos Acedo Mendoza. *Venezuela: Ruta y Destino*, 1971, are analyses of contemporary social problems of Venezuela.

8. Federico Brito Figueroa. *Venezuela Siglo XX*, 1967; Domingo Alberto Rangel. *Capital y Desarrollo*, consisting of three volumes: *La Venezuela Agraria*, 1969, *El Rey Petróleo*, 1970, and *La Oligarquía del Dinero*, 1972.

9. Federico Brito Figueroa. *Venezuela Contemporáneo: País Colonial?*, 1972, and other works by the same author reflect a modern Marxist point of view.

10. W. W. Rostow. *The Process of Economic Growth*, New York: Norton, 1962; *The Stages of Economic Growth*, Cambridge: Cambridge University Press, 1960.

11. Alfredo Machado Gómez. *Crisis y Recuperación*, 1972, pp. 249–284.

12. John Friedman. *Regional Development Policy: A Case Study of Venezuela*, 1966, pp. 127–131.

13. W. W. Rostow. *The Process of Economic Growth*, 1962, p. 289.

14. "Economic Developments in Venezuela in the 1950s," *Economic Bulletin for Latin America*, Vol. V, No. 1, March 1960, p. 21.

15. Douglas North. "Location Theory and Regional Economic Growth," *Journal of Political Economy*, Vol. 63, June 1955; Harold Innes. *Problems of Staple Production in Canada*, 1933.

16. Colin Clark. *Conditions of Economic Progress*, 1940.

17. The dependence literature is large. Two samples, with special reference to Latin America, are André Gunder Frank, James D. Cockroft, and Dale L. Johnson. *Economía Política del Subdessarrollo en Amerérica Latina*, Ediciones Signos, Buenos Aires, 1970; and André Gunder Frank. *Capitalismo y Subdesarrollo en América Latina*, Ediciones Signos, Buenos Aires, 1970. Dependence is, of course, related to Marxist analysis; Luís E. Aguilar. *Marxism in Latin America*, New York: Knopf, 1968, contains articles dealing with dependence.

18. Armando Córdoba. *Inversiones Extranjeras y Subdesarrollo*, 1973, p. 222.

Chapter II

From Pearls to Petroleum

From the beginning and throughout nearly five centuries of history the swirl of international economic and political systems caught Venezuela up and tossed it about, never quite letting it go. Venezuela was never fully in command of its own fate. Economic and political forces from the outside have dominated its economy since the discovery. Other country's policies and international markets have always dictated specialization in one or a few products; the principal feature of its productive system has been the series of specializations which dominated the economy from 1498 to the present.

Domingo Alberto Rangel put it eloquently:

> From the first moment that America was born it was charged with the responsibility of providing exotic goods for the sumptuary consumption of the directing classes of Europe and precious metals to lubricate its monetary circulation and to facilitate its capital accumulation . . . America would produce those products which Europe needed to fulfill its historic role. America would produce through those institutional systems implanted by the conquerors. And finally, America would produce for the benefit of Spain.[1]

Cubagua and Conquest

Venezuela was the first landfall of the Spaniards on the South American continent. On his third voyage to the new world in August 1498, Columbus made contact with Indians on the mainland of the Peninsula of Paria. Subsequent voyages explored the Caribbean coast from the mouths of the Orinoco River to Cartegena, west of Lake Maracaibo. On observing villages built on stilts in the lake, Américo Vespuccio named the area little Venice, or Venezuela.

At the time of the first landing about 350,000 Indians, divided into many tribes and nations, scratched out a subsistence by primitive agriculture, hunting and fishing. The valleys of the Andes and Coastal ranges, around Lake Maracaibo, and along the major rivers were the centers of this Indian civilization. The Indians were sedentary, living in small communities, and producing corn and yucca. Small plots and slash-and-burn agriculture predominated.

Some of the more advanced tribes, such as those in the Andes, terraced the mountain slopes and even irrigated their crops. The Indians had a limited exchange economy and some tribes used primitive money. The northern tribes, the Caribs, were warlike. The Indians were in permanent possession of the productive land and made use of the land in diversified agriculture. They had not amassed any significant wealth and had no great civilization, as was the case of the Aztecs in Mexico and the Incas in Peru.[2]

Cubagua is a barren rock in the Caribbean Sea off the coast of Venezuela. It was the site of the first exploitation of Venezuela's natural resources by the Spaniards. Columbus, on his first trip, observed that the Indians wore necklaces of pearls. While known in Europe, the pearls were still an uncommon adornment and much prized. Columbus reported his observations and the news spread rapidly. The first voyage to Cubagua returned to Spain with 80 pounds of pearls in sacks. In 1500 Spanish adventures established a settlement, the first settlement in what is now Venezuela.[3] In 1530, for example, 800,000 duros (Spanish dollars) in pearls arrived in Spain; the value of mining operations in Spanish America in that year was 2,000,000 duros.[4] The early exploitation of the pearl beds almost destroyed them.

Spaniards also initiated the conquest and pacification of the mainland while Cubagua was going through its life cycle, which ended in a disastrous earthquake. Expeditions went forth from Cubagua in search of gold, pearls and slaves. They explored the interior of the mainland and established communities which the Indians, now familiar with Spanish ways and increasingly hostile, destroyed. Missionaries fared little better, often paying with their lives for the misbehavior of the conquerors. Even the celebrated Fray Bartholomé de las Casas, apostle of the Indies, was unsuccessful in his bid to promote the humane settlement of eastern Venezuela.

The Spanish king owed the German banking house of Welser for helping him become the Holy Roman Emperor. In 1528 he leased western Venezuela to the Welsers for exploration and town-building. The El Dorado myth was strong. The Germans scoured the interior for the fabled city and its gold-covered king. Not finding it only increased their appetite for lesser forms of wealth. They forgot town-building and plundered the land with the help of Spanish soldiers. Finally, in 1546 the Spanish king revoked the lease.[5]

The second half of the sixteenth century witnessed a continuation of the conquest. In bloody engagements the Spanish asserted their superior arms and civilization, and the land fell into their hands. The greatest ally of Spanish arms was European disease, which decimated the population. But even while conquering the Indian and destroying his civilization, the Spaniard mingled blood with the Indian woman. The Venezuelan, part Indian, part white, was the result.

By the middle of the sixteenth century, even before the conquest was complete, the Spaniards had assembled a general knowledge of the geography and resources of the territory. Great wealth did not exist. The El Dorado myth dissolved. If the Spaniard were to receive any benefit, it must result from Indian labor working the land and producing a surplus. Hence, the conquerors devised methods to use the Indian and usurp his land. The principal methods were the construction of communities and the encomienda.

The Encomienda

The Spaniards used the encomienda to marshal the two factors of production—land and labor—for the benefit of the peninsular Spaniard and the tiny minority in America. Columbus initiated the system on Hispaniola shortly after discovery, but for years it had no royal sanction. The encomienda spread through Spanish America and was the predominant settlement pattern in the sixteenth and seventeenth centuries. Royal decrees legitimized the practice and by 1520 the encomienda was well established. Because of the slow conquest of the province, the earliest encomienda in Venezuela did not begin until 1532.[6]

In return for the benefits of Christianity and civilization the Indians owed the encomendero a certain amount of work, a specified number of working days or an amount of produce. The encomendero determined the nature of the work and supervised it. The Indian was free to work on his own account and grow crops on a small plot when he had fulfilled his obligation to the encomendero. The Indian was not a slave but he lacked freedom of movement and independence of action. The encomienda tied him to a specific piece of land and to its owner for life.

The Indian received benefits from the encomendero. The encomendero had to protect the Indian, provide religious instruction, assure him justice, and provide the local governmental and legal framework within which the Indian lived and worked. The Indian cacique, or chief, was the intermediary through whom the Spaniards imposed their rules and behavior patterns, which included life in a Spanish community. The encomienda usually expired when the encomendero died, but on occasion the right existed through the life of one or two heirs. The land was a royal grant and passed to heirs even if the encomienda ended. The Indian, however, often remained and worked for the property owner but without the rules and obligations of the encomienda.

The economic function of the encomienda was to produce an agricultural surplus. Land under the Indian yielded a meagre surplus, often none. Productivity was low and wealth accumulation was not a strong cultural trait of the Indian. Agricultural production was most often food crops for immediate or near-term consumption. The Spaniards demanded food, clothing and shelter, as well as their customary goods and services, according to European standards. In addition, they wished to amass wealth in order to return home as great lords. All this was far beyond anything the Indian could provide. The royal land grant and the encomienda helped to increase productivity and provide the surplus.

The encomienda also initiated agricultural specialization and exchange, including overseas trade. By altering land and labor use the encomienda paved the way for the tobacco and hide specializations of the seventeenth century, cacao in the eighteenth, and coffee in the nineteenth and twentieth centuries. Specialization began to take hold by the end of the sixteenth century. In 1599 flour was the most important export. By 1607 tobacco was in first place, flour second, and hides third. While exports were small they demonstrate how the colonial agricultural system, the key institution of which was the encomienda, began at an early date to turn Venezuela into a surplus agricultural province.[7]

The encomienda did not survive. The crown abolished it in Venezuela in 1687, although vestiges lingered into the eighteenth century. In its 155 years, however, it determined in large measure the land use and tenure system, much of which lasts to this day. The great hacienda emerged from the land grants to the encomenderos and the labor force was the Indian left by the encomienda. The Indian became the campesino who worked not only on the hacienda but also on his small plot.

Tobacco and Cattle

Tobacco and hides were the first two agricultural products to achieve prominence in Venezuelan exports. While exports of flour, cotton cloth and other products began in the middle of the sixteenth century, they were unimportant both to the province and to Spain. Tobacco was first in exports in the early part of the seventeenth century, followed by hides which had their turn for more than half a century. Tobacco returned to the lead in the last quarter of the seventeenth century. For a while Barinas tobacco achieved some fame in Europe. Finally, at the turn of the eighteenth century, cacao overshadowed both tobacco and hides. While exports were important in the first two centuries, most agricultural production was for the domestic population. The most important crop was corn.[8]

The pre-Colombian Indian did not have cattle or the horse. The Spaniards introduced them early in the conquest. By 1560 cattle began to populate the llanos, a natural grazing and breeding zone. The horse became the principal means of transportation and beef a food staple. Cattle often ran wild in the llanos, living off the land. The hide trade began at the end of the sixteenth century. By 1622 exports were 13,500 hides. Great variations year to year in exports occurred, ranging from a few thousand to 24,000 hides. Between 1620 and 1680, hides accounted for three-fourths of exports.

The star of tobacco also shined brightly in the seventeenth century. Columbus discovered native plantings and under the Spaniards tobacco became a commercial crop. At the beginning of the seventeenth century its export exceeded other products, but the king forbade its planting in the coastal region, in order to stop contraband trade with Curacao. It dropped into second place but emerged again as an export late in the century, only to bow to Virginia tobaccos at the end of the century.

Cacao

The eighteenth century was the epoch of cacao. The tree is native to Venezuela and the country possessed a productive zone of high quality cacao land through the Coastal range. Much of the cacao went to Veracruz and smaller amounts directly to Spain and the Canaries. By 1631 there were one-half million trees and exports had climbed to 220,000 pounds. Exports continued to grow during the century, averaging after mid-century more than 1,200,000 pounds per year. At the beginning of the eighteenth century Venezuela for the first time showed real promise of paying off as a colony through cacao exports.[9]

The war of Spanish succession following the death of Charles II in 1700, led to depressed conditions and temporarily reduced exports. Ship communication was rare and the Venezuelan economy was in a slump. When the war was over, Spain, under the Bourbons, encouraged trade and began to experiment with monopolistic trading companies, in use by other colonial powers. In the early part of the century Spain tried small trading companies and considered others in an effort to reduce privateering and contraband, and to encourage exports. These experiments led to the establishment of the Compañía Guipuzcoana, better known as the Caracas Company, in the second quarter of the century.[10]

A royal order established the company in 1728. Its owners included the king, the merchants of Guipúzcoa, a northern Basque province of Spain, and wealthy landowners in Caracas. It had a complete monopoly of Spanish trade with Venezuela. Its task was to supply Venezuela's import needs and to carry cacao to Spain. It was also to provide for the internal and external security of the colony, to eliminate pirates and contraband trade, and to promote agriculture. In short, the Caracas Company was to make Venezuela safe and profitable for Spain.

The company contracted to send two armed ships each year. The first ship sailed from Spain in 1730 and on arrival it set up security and trading services, sailing back with 1,500,000 pounds of cacao. The company made a profit of 450 percent on the transaction. Before 1730 Mexico had imported nearly all Venezuelan cacao. Under the company, much of the cacao went directly to Spain. During its existence the company handled about one-half of all Venezuelan cacao at prices set by the company. Profits came from the import side as well, since the company was the only importer, wholesaler, and the most important retailer.

The company became unpopular quickly. The principal benefits went to the crown and Spanish merchants, not to Venezuelans, and the company paid no taxes. Anti-Spanish sentiment was strong and Venezuelans resented the rapidly acquired wealth and success of the Basques. Elimination of contraband trade, profitable to Venezuelans, fueled the controversy. Two rebellions broke out and while neither succeeded, the whole atmosphere of Venezuelan-Spanish relations were soured by the company's behavior and power. Despite economic prosperity and growing cacao exports discontent was rampant and conflict frequent. In 1781 the king abolished the monopoly and Spain, sensing the gathering storm, liberalized trade.

Independence

By 1800 Venezuela and Venezuelans had come into existence. In the three colonial centuries population had little more than doubled, but its composition had changed radically. A small Spanish ruling group possessed political and military power. A fourth of the population was white criollo of Spanish descent but born to the colony. They had substantial economic power but were subject to the ruling peninsulars. About 60,000 black slaves had arrived from Africa to work on cacao estates. Some unassimilated Indians remained. The most numerous group was the pardo or mestizo, a mixture of Indian and white.

Change was in the air. Spain controlled the treasury, trade and the military forces; these were her instruments of power. The recession at the end of the eighteenth century had dissipated but the colonies were not prosperous. The colonial system had passed the zenith of usefulness as an economic organization. Large estates in the hands of absentee owners, slavery, a complex system of royal taxation and trading, the problem of contraband trade despite Spanish efforts to suppress it, dependence on cacao, and class and racial antagonism and fear all pointed to a change in the making. The criollos especially were restive under Spanish rule.[11]

The Spanish colonial system was unstable; the successful North American revolution and the French revolution shook it badly. Revolutionary ideas were blowing throughout the empire. The Caracas Company had created a maelstrom of discontent in Venezuela through its imperious behavior and had unwisely permitted the importation of books. Tom Paine and Ben Franklin were popular authors in Caracas. On a more mundane level, the criollos were tired of sharing the largess of their land with Spain and began to toy with the idea of economic reform or perhaps even independence.

The excuse for the onset of the independence movement was French adventurism in Spain. Napoleon invaded and occupied Spain, captured Fernando VII, installed his own brother on the throne, and initiated the last phase of his brilliant rise and fall. At first, and for a brief moment in 1810, Venezuela remained loyal to Fernando and repudiated only the French puppet government. But hardly a year passed before the Venezuelan criollos insisted on complete independence.

The result was a cruel and destructive war of independence that lasted for eleven years on Venezuelan soil. Destruction was everywhere; tens of thousands died. The ebb and flow of battle across the valley floors destroyed crops; owner and campesino went off to war and the hacienda deteriorated. The economy ground to a halt with misery and chaos everywhere. Production and exports plunged, not to recover for twenty-five years. Exports in 1831–1832 were less than 60 percent of the 1810 level. Cotton exports almost disappeared; cacao exports were halved; indigo, hides and cattle were 10 to 25 percent of their former levels. Only coffee exports increased, by 43 percent.[12]

The impact of the war was severe. The cacao trade, almost stopped by the war, recovered slowly. In the nineteenth century it dropped into second place as an export and retained that position. The destruction of crops, properties and life was the most damaging effect of the war. The loss of life among the criollos was catastrophic. Indeed, Venezuela lost a substantial part of its land-owning class during the war; the institution of latifundio tottered on its foundations but did not topple. A new directing class emerged from among the revolutionary military and took over the properties.

Venezuela achieved political independence and nationhood. The empire disintegrated and balkanized. Economic independence did not accompany the political independence. The nature of the economic dependence, however did change. Before 1821 most economic decisions for Venezuela originated in Madrid, Sevilla and Cádiz, by kings, merchants and officials. After independence Venezuela entered the swirl of world capitalism, to be tossed about by markets dominated by European middlemen. The weight of the new republic was negligible in those markets. Venezuela produced two tropical crops in increasing demand in Europe, but Venezuela was one among many producers.

The coffee and cacao markets were highly competitive. The individual producer sold to middlemen, representatives of European buying houses, resident in Venezuela. The middleman collected and shipped the products to his home office which processed and then sold the products on the final goods markets. No one controlled the final goods market; it was impersonal and largely immune to manipulation by individual producers or by consumers. At the intermediate stage, however, a few great houses and their agents abroad did exercise influence over producers, even though the middlemen themselves were subject to market influences. The middleman in any producing area was the only, or one of few buyers from producers. Technically the producer had the option of shopping among middlemen but this seldom happened in practice.

The planter himself had no market connections and tended to avoid uncertainty by selling his output year after year to the same middleman. A producer, by custom, risk avoidance and propinquity came to depend on a single middleman, who therefore influenced the price at which he bought. The middleman was also often the source of capital, directly or indirectly, which the producer needed not only to initiate but also to sustain or expand operations, solidifying the influence of middleman over producer.

The Rise of Coffee

Coffee became the kingpin of the Venezuelan economy, and its dominance lasted for a century. The rise of coffee, already a major crop when the independence movement began, accelerated with the war. The war resulted in structural changes with respect to the availability of land, labor and capital. The new structure, in conjunction with changes in product markets favoring coffee, made coffee cultivation increasingly attractive to the Venezuelan planter. Cultivation of coffee was quite limited until the end of the eighteenth century. In 1800, 6,600,000 libras (a little more than a pound) were exported. Venezuela exported similar quantities in the years preceding the outbreak of hostilities.

A series of interrelated factors was responsible for the changeover from cacao to coffee. One element was time. When the planters could get back to their crops after the war, they were under pressure to begin production and start earning income as quickly as possible. They returned those trees not destroyed or damaged by war to production at once. But the cacao valley floors were the battlefields, and destruction was extensive. Rather than face the six-year gestation period for new cacao trees, many planters opted for the three years required to bring coffee into production.

The colonial cacao markets, Mexico and Spain, were cut off during the war and were either gone or uncertain in the postwar period. Europe, during the unsettlement of the first part of the nineteenth century, had sought out new producing areas. In addition, the relative prices of coffee and cacao had changed so that by the end of the first quarter of the century the market prospects for coffee appeared brighter than for cacao. A given piece of land, suitable for either crop, yielded twice as much income if planted to coffee. This reflected the booming demand for coffee in Europe. The high price did not endure, but coming at the end of the war and in the early postwar period, high coffee demand did much to encourage the shift to coffee and to commit the land to its production.[13]

The land, furthermore, was available in Venezuela. Some of the land came from cacao and other crop lands laid waste by the war and not replanted to the same crop after the war. Still more land came from the slopes of the valleys not suitable for cacao but usable for coffee. Planters cleared and planted this readily available and low cost land to coffee trees. New young trees and virgin soil yielded high productivity.

Labor posed few problems. Despite many casualties the returning soldiers, population growth, and the labor released from other crop production were sufficient to supply the needs of the new crop. The manpower demands of coffee, furthermore, were not as great as for cacao, and subsistence agriculture tended to be overpopulated. The seasonal nature of coffee culture permitted drawing from subsistence agriculture in season, releasing these workers for the rest of the year.

Capital was the most serious problem. Before independence the church, Spanish capitalists, and a few wealthy criollos provided the capital for cacao and coffee production. The war destroyed all three sources of credit. The coffee expansion had to depend on foreign credit sources. The credit came from North American and European speculators and middlemen who were observing the rise in coffee prices. Thus, capital became available through the consumers of coffee in the early period.

The cost of credit was high. In the late 1820s the rate of interest was two or three percent per month, but it declined to one percent per month in a decade. In addition, while the church and other pre-war lenders provided long-term loans, after the war only short-term credit was available from foreigners, commission houses and middlemen. The terms of credit became a major issue dividing the lending and borrowing sectors of the economy.

As Venezuela emerged from the war, political control fell into the hands of urban financial and commercial interests, a coalition of political and economic groups which formed the Conservative party, espousing the laissez-faire economic policy. The government relaxed the traditional guarantees to the borrower, allowing any rate of interest and establishing a special tribunal to handle debtor cases. Credit became available on favorable terms to lenders, who were in power, but on onerous terms to borrowers who formed the Liberal party in order to gain political power and reverse the credit policy.[14]

Coffee Prosperity and Depression

In 1829–1830 coffee production was 2,882 metric tons, having climbed from a much lower point at the end of hostilities in 1821. The 18-year period from 1824 to 1842 was an almost uninterrupted expansion, reaching a peak of 15,180 metric tons. In addition, for much of this period prices also rose. For example, in 1830–1831 the price was Bs620 per ton; in 1841–1842 it was Bs1,100 per ton. Small ups and downs did not obscure the pronounced trend toward accelerating prices and production. It was a feverish time in the coffee sector.[15]

Toward the end of the 1830s cracks began to appear in the coffee expansion. In 1839–1840 production dropped slightly but recovered the next year. Production was beginning catch up with market demand as new producing areas entered the market. In addition, Venezuelan planters were in a squeeze. Increasing production began to put pressure on land availability and land costs began to rise. Labor was also beginning to run short. As production increased, more and more workers had to labor among the trees year-round, and seasonal harvesting made huge demands on the labor force.

The most critical problem was still capital. Planters entered the vortex of an upward spiral of expansion and increasing indebtedness, all predicated on still more exports at the same or higher prices and on the continued availability of credit. The credit, however, was short-term and high cost. As production costs, such as land and labor costs rose, a turn in the market rendered the planter unable to pay off his debts or to renegotiate his loans. Creditors called outstanding loans and took over the property if the planter was unable to pay.

The house of cards of coffee collapsed. Exports and export prices declined. Prices had declined slightly in the late 1830s. Hoping for a recovery of prices, planters borrowed more and more, thus deepening their crisis. When it was apparent that the market had turned, financial interests began to protect themselves, setting the stage for an economic conflict between financial and agricultural sectors. The conflict soon turned into a political conflict between the Conservatives and the Liberals.

The Conservatives blamed the mismanagement of planters. The money, they claimed, was loaned in good faith and if the planters could not compete then they must give way to those who can. The solution was: belt-tightening, the open credit law which would attract capital, and sound long-term public works. The Liberals saw it otherwise. Greedy capitalists and the credit law were the culprits. The government must protect borrowers and eliminate the pernicious law. The government must also make easy credit available to agriculture, since it was the backbone of the economy.

As the recession worsened in 1843 and 1844 the conflict became increasingly sharp. Civil disturbances were common. The Conservatives retained political power but faced mounting unrest and discontent. In 1846 a revolution erupted but the government squelched it and the Conservatives won the presidential election. Their candidate, the new president, however, turned Liberal after acquiring office and enacted much of the Liberal program, including the repeal of the open credit law. Since the government had no money, no easy credit was forthcoming.[16]

Coffee production recovered but began to experience wide year-to-year variations. Prices did not recover. Coffee planters remained in straitened conditions, and although they had acquired political power, economic recovery did not take place. This time, it was the turn of the Conservatives to rebel. Civil unrest and uprisings were again commonplace and the political and economic state of the nation at mid-century deteriorated to chaos. What had begun as an economic issue became a political conflict which then led a life of its own, worsening still further the economic problems, culminating finally in the Federalist War.

The coffee crisis of 1842 and the ensuing period illustrate two characteristics of Venezuelan economic development in force for more than a century. First, it was a turn of the market (over which the country had no control) which initiated the crisis. Venezuela depended upon an uncertain and changing market. While the behavior of planters and the government might have mitigated the crisis, dependence on external markets was the underlying cause.

Secondly, no compromise among the contending parties was possible. Positions on economic policy defined the political stand of both parties and neither could accept an intermediate position. Each felt the only way was its way. The opposition had to be defeated, by force if necessary. The result was an incomplete and protracted recovery. These two characteristics persisted until the middle of the twentieth century.

Coffee Dominance

At mid-century, production and exports were at their highest levels but prices were still down. The 15,000-ton high of 1841–1842 gave way systematically in the 1850s and 1860s to higher production, although internal strife and civil war disrupted production. Cacao exports also suffered and even declined in the 1860s. Despite the unrest and uncertainties coffee continued to dominate the economy and exports grew. The international economy pulled the Venezuelan economy along.

The last thirty years of the century were relatively prosperous. The 1870s ushered in an era of political stability and economic growth based on coffee expansion. In 1872–1873 exports were 34,273 tons, nearly double the previous highs. Unit export value was Bsl,250 per ton, also higher than previously. Through the decade, exports varied year-to-year but the value of exports rose because of rising prices. The 1880s were also prosperous, with exports varying from 39,000 to 51,000 tons. Prices declined from the earlier highs but export receipts were still on a rising trend. The 1890s were the apogee of coffee exports, averaging more than 50,000 tons a year. Prices, however, declined. In spite of lower prices the value of exports continued high, except at the end of the decade when prices plunged to Bs720 a ton. The falling prices at the century's end stirred up unrest.[17]

The site of coffee production in the national period was the Andean region of western Venezuela. The andinos, however, had participated only marginally in the political life of the country. The country was in the hands of caudillos from the llanos and Caracas. The price declines of the 1890s, however, brought the andinos out of their isolation. A golpe de estado in 1899 placed the andinos in power, a position they held firmly until 1945 and then relinquished only briefly to rule again from 1948 to 1958.[18]

Despite the andino rule, coffee never again achieved the dominant position it held in the last quarter of the nineteenth century. By the turn of the century Venezuela had run out of good coffee land, and labor was increasingly scarce and high cost. The industry was growing old and productivity declined. New producers, notably Colombia and Brazil, had displaced Venezuela as the major coffee suppliers. By 1900 Brazil was overwhelmingly the principal producer. Venezuelan exports did not decline in the early years of the twentieth century but prices were low and it was no longer possible to expand production significantly.

European unrest and the first world war stimulated prices and exports, providing momentary respite from the coffee decline. But in 1926, even though coffee exchange earnings were still high, coffee slipped silently into second place behind petroleum. From 1825 to 1926 coffee had ruled Venezuela, and the international coffee market, in which Venezuela had no say, had ruled coffee. The economy grew in function of North American and European demand, and coffee middlemen dominated the producers. The market rewarded and punished Venezuela and permitted only slow and uncertain growth.

It is too simple to say that cacao in the eighteenth and coffee in the nineteenth century were all of the Venezuelan economy. They were the leading sectors, the principal exports, and among the most important economic activities. But Venezuela grew other crops for export. Hides were important, as were cotton, indigo, tobacco. In addition, the country produced most of its food. Agriculture was not efficient and subsistence agriculture supported a large part of the population. The land tenure system—simultaneous latifundio and minifundio—provided ample disincentives. Only minor manufacturing existed, transportation and other infrastructure were deficient, and the service sector, including banking and commerce, was anemic.

Early Petroleum Developments

The Indians knew of the presence of petroleum and used it for medicinal purposes. The early conquerors commented on the asphalt beds and oil seepages. Indeed, in 1539 Venezuela exported its first two barrels to help cure the king's gout. Pirates and seamen used the asphalt to caulk their ships. It was also used for lighting. But the resource was devoid of economic value until late in the nineteenth century when inventors in the industrial countries discovered how to convert petroleum into mechanical and electrical energy.

A happy circumstance resulted in state ownership of all subsoil resources in Venezuela. As early as 1559 Felipe II incorporated all mines in Spanish America into the royal patrimony. Later Spanish monarchs reaffirmed the rule. By the time of independence it was a well-established principle that the owners of the land had surface rights only and that the state owned the sub-surface. In 1829, following the Spanish tradition, Simón Bolívar decreed that mines of all types were national property. Throughout the nineteenth and twentieth centuries successive governments have confirmed this national ownership.[19]

In advanced countries petroleum production on a commercial scale began. In 1858 in Canada and in 1859 in the United States production began. The multiple uses of petroleum for illumination, lubrication and as a source of energy burst upon the world; an industry was born. Rockefeller bought up many small North American refineries in the 1870s and established Standard Oil in 1882. In Europe in 1890, Royal Dutch boomed. The petroleum age was dawning.

Back in Venezuela, Manuel Pulido formed a small company in 1872, the Compañía Nacional Petrolia de Táchira, on a small concession near San Cristóbal. An earthquake the year before had opened up a fissure. Pulido bought a drilling rig but it broke up in trying to get it over the Andes. So he dug a shallow well by hand, built a small refinery, and supplied the region with kerosene and gasoline. It was the first oil production and refining in Venezuela. The company never became large but it was successful and indeed lasted until 1935.

Toward the end of the nineteenth century the concessions for asphalt and petroleum multiplied. They were of little importance since the concessionaires had few resources and the purpose was to make money off the concessions. Most lapsed without any exploitation. Yet at the same time, the geologists were making a serious study of the Venezuelan subsurface. The petroleum industry was still in its formative stage, lacking in capital, technology and markets.

In 1905 a new law laid down the terms of concessions. In 1909 British interests, through John Allen Tragelles, received a large concession. This later became the property of Caribbean Petroleum. In 1910 General Asphalt received a concession and on it the first real oil well was drilled in 1912. Royal Dutch Shell later acquired both General Asphalt and Caribbean Petroleum concessions. Still other concessions fell into Shell hands later. In 1914 the first field came in. In the next few years many new discoveries turned up. In 1917 petroleum appeared for the first time in the export statistics of Venezuela.[20]

The Last of the Caudillos

In 1899 an army of sixty men had invaded Venezuela from Colombia and in a comic opera war had marched to Caracas and seized power. General Juan Vicente Gómez was second in command of that army and for nine years he served his chief faithfully. When Cipriano Castro left the country in 1908 General Gómez took command and initiated the longest and harshest dictatorship Latin America had ever witnessed. His regime was so cruel that no street nor monument nor plaza honors his name even today, and he lies in a nearly forgotten grave. In a negative way, however, he was a major contributor to the economic and political developments which began with his death in 1935.

General Gómez was a farmer and rancher. Venezuela was his hacienda, not only figuratively, but, as his political dominance became absolute, almost literally. He acquired most of the best properties in the country, often by less than honorable means. He was a careful and competent rancher and since the Venezuelan government was a part of his hacienda, he improved it; he unified and modernized the government. He established centralized institutions. His national army crushed the regional caudillos and extirpated decentralized political power. He eliminated the nineteenth century political system. His efficient spy system cowed the political opposition and his prisons hosted the best of Venezuela. His national treasury created prosperity for the private and public elite and produced povery and misery for most Venezuelans.[21]

General Gómez was uncannily astute and absolutely ruthless. Even his milder forms of punishment destroyed men's minds and wrecked their bodies. If he imprisoned a friend by mistake, he stayed in prison. General Gómez knew that friendship would not survive a stint in his prison. He gave Venezuela a taste of what a really efficient dictator is like and Venezuelans learned to loath and fear him. Without that hard and long lesson, Venezuelans might now be less tolerant of some of the inefficiencies and abuses of democracy.

General Gómez never understood oil. He did understand money and he viewed petroleum as simply another way to enrich himself. Following the promising geological studies before the first world war, Venezuela soon became an important exporter. The arrival of foreign capital was an extraordinary opportunity for the general to consolidate his power. He welcomed the companies and created conditions which the foreigners found attractive. His modest financial demands enticed the companies, but were large compared to previous revenues. He took the money and let the companies do as they pleased.

The oil companies, in cooperation with General Gómez, exploited Venezuela. The conditions imposed by the dictator greatly favored the companies. General Gómez also empowered the companies to establish many of their own conditions, including writing the country's petroleum laws, which included a remarkable absence of supervision by the government. It is not surprising that the foreign companies came in force and came to stay, making a huge investment in the country. More stringent conditions might have benefitted the dictator more, but not necessarily the country, and such conditions might have deterred the massive entry of foreign capital. Unintentionally, the leniency established an industry which at a later date could be deployed in the national interest.

The Venezuelan economy experienced modest economic growth during the Gómez regime. The petroleum industry grew rapidly but low taxes, small demand for domestic labor, and limited linkages with domestic production confined the benefits to foreign owners and their clients. Petroleum drew capital and labor from agriculture to the campsites and damaged agriculture. Petroleum also provided sufficient foreign exchange so that domestic agricultural and industrial producers could not compete with imported goods. In addition, the dictator and his supporters siphoned off into their personal fortunes the meager economic gains which the economy produced.[22]

Petroleum Captures the Economy

As production and exports of petroleum grew there was frenetic concession activity. British interests were primary until 1919. Then the North American entered the lists, obtained important concessions, and started exploration. Companies used their legations to influence the government which shamelessly sold concessions for the personal profit of officials. Even the chief executive engaged in a series of concession transactions for his own profit. Chicanery and influence peddling were rampant. Royalties and taxes were low and no income tax existed. The government imposed minimum regulations, often those specified by the companies. Production, however, began to flow and climbed at a dizzying rate.

In 1928 Venezuela was the largest petroleum exporter in the world and the second largest producer after the United States. Between 1925 and 1929 exports increased almost seven times and foreign exchange receipts more than doubled. The fiscal income of the government went from Bs21 million to Bs51 million. In 1925 Venezuela exported two percent of the world total; in 1919, nine percent. The petroleum sector was not quite 10 percent of total product in 1925; in 1929 it was more than 29 percent. In 1925 petroleum exports were 28 percent of total exports and the corresponding figure for 1929 was 45 percent and rising each year.[23]

The dependence of Venezuela on petroleum was almost immediate and complete. It was manifest not only in the statistics of trade and production, but also in decision-making. The government had no voice, beyond the granting of concessions, in the industry. Decisions as to when and where to explore, build roads, drill wells; decisions on rates of extraction and exportation; rates of investment, personnel and labor policies, and many other decisions were exclusively in foreign hands. The companies did not consult or even tell the government. They wrote the petroleum laws, made and unmade ministers. Only in a technical sense was Venezuela a sovereign state with respect to oil.

The petroleum industry convulsed the domestic economy. The goal of the companies was to make money by feeding the hungry and growing market in Europe and the United States. They did it without regard to the local economy and in the absence of policies to protect Venezuelan nationals or their interests. Domestic production could not meet the demands of the new industry and its employees. Imports increased rapidly.

The on-site Venezuelan labor force was limited, so the oil companies imported people. Even so, the expansion of petroleum induced a massive population movement from the countryside and small towns to Maracaibo and other producing regions. Not finding sufficient work in the oil fields, the Venezuelan labor force occupied itself in marginal commercial and service activities and often remained unemployed.

The withdrawal of the labor force from agriculture and other sectors had an adverse effect on production and costs. Often the most productive workers left, and labor costs rose. Agriculture became less productive and less profitable and domestic capital fled to commerce and support activities near the petroleum fields and communities.

The flow of oil did increase incomes somewhat, and these larger incomes plus expectations of even larger incomes had the effect of increasing the birth rate. The death rate also declined as public expenditures were pumped into health measures. Population began to grow, with one break in the trend line occurring in 1920, another in 1935.

The worldwide depression hurt exports but the damage was neither serious nor of long duration. Exports dipped from 20 million tons in 1930 to 17 million tons for the three succeeding years, but then increased continuously. Export receipts did not recover as rapidly since prices remained low. Government income was sufficient enough that General Gómez paid off the government debt completely in the middle of the depression.

Venezuela in 1935

Despite the trappings of a modern democracy—a written constitution, Congress, judiciary, checks and balances—Venezuela was the personal fiefdom of General Gómez until he died on December 17, 1935, 110 years to the day after Simón Bolívar died. In the last years the military challenges had disappeared and the student revolt of 1928 petered out. General Gómez had silenced the opposition with the grave, jail, or exile, and his supporters occupied the Congress, the bureaucracy, and the economic oligarchy. His spy system intimidated even the latent opposition.

The world was in depression. The price of petroleum was declining and even though in his last year production had recovered, the Venezuelan economy was stagnating. The depression had hit agriculture hard; coffee and cacao were in the doldrums. General Gómez had paid off the domestic and foreign debt and maintained a balanced budget. Domestic liquidity was scarce, investors cautious, and commerce slow. Even so, the depression in Venezuela was less severe than in many countries because the drop in petroleum was less steep and recovery more rapid.

National income per person was about $150, but the distribution of income was so inequitable that most Venezuelans had only a bare subsistence. More than three-fourths of the 3.3 million people still lived in rural areas where subsistence agriculture fended off starvation. The cities were commercial centers with no industry. Roads, communications and public services were primitive and not improving. Venezuela had the highest incidence of malaria in the world. Education touched only a few. In 1935 Venezuela was poor and economically and politically backward.

Public policy served the interests of the dictator exclusively. The goals of the regime were security, political stability, and the enrichment of the dictator, his family and his supporters. Petroleum served to obtain revenue sufficient to the government's needs without discouraging private foreign capital. The regime repulsed efforts of the companies to establish refineries, fearing labor and political problems which might result from industrialization and the revival of regionalism in the state of Zulia, site of the oil fields.

A balanced budget or a surplus was the rule. A private bank handled the government's funds and no monetary policy existed. Government regulation and intervention in the economy served the regime. No questions of labor unions, social security, capital markets, trade associations, or income distribution marred the tranquility of the dictator's last year. Economic development was not in the lexicon of Venezuela. Democracy was a dream of exiles.

NOTES

1. Domingo Alberto Rangel. *La Venezuela Agraria*, 1969, pp. 13–14.

2. José de Oviedo y Baños. *Historia de la conquista y población de la provincia de Venezuela,* 1824, 1967, is a classic treatment of the early conquest.

3. José Gil Fortoul. *Historia constitucional de Venezuela*, 5th edition, Vol. 1, 1967, p. 35.

4. Tomás Polanco Martínez. *Esbozo sobre historia económica venezolana*, 2nd edition, Vol. 1, 1960, pp. 44.

5. Juan Friede. *Los Welser en la conquista de Venezuela*, 1961, is a study based on recent scholarship and tends to be more generous to the Welsers than earlier works which harshly condemn their activities.

6. Eduardo Arcila Farías. *El regímen de la encomienda en Venzuela*, 1966, leaves unresolved the controversy surrounding Spanish treatment of Indians and the effect of the encomienda. This volume, by one of the leading Venezuelan historians of the colonial period, deals with the encomienda in Venezuela and property rights.

7. Eduardo Arcila Farías. *Economía colonial de Venezuela*, 2nd edition, Vol. 1, 1973, pp. 112–113.

8. Eduardo Arcila Farías. *Op. cit.*, chapter IV.

9. Eduardo Arcila Farías. *Op. cit.*, chapter V. Dr. Arcila Farías has laboriously compiled export data from the entire seventeenth and eighteenth centuries.

10. Roland Dennis Hussey. *The Caracas Company, 1728–1784*, 1934, is still the standard authority on the Company.

11. Manuel Vicente Magallanes. *Historia política de Venezuela*, Vol. 1, 1972, deals with events leading up to the independence movement.

12. Carlos A. D'Ascoli. *Esquema histórico económico de Venezuela*, 1970, p. 247.

13. John V. Lombardi. *The Decline and Abolition of Negro Slavery in Venezuela 1820–1854*, 1971, p. 97, and statistical sources cited in footnote 15 below.

14. José Gil Fortoul. *Op. cit.*, Vol. 1, chapter III; and Tomás Polanco Martínez. *Op. cit.*, Vol. 2, chapters IV and V.

15. Ramón Veloz. *Economía y finanzas de Venezuela desde 1830 hasta 1944*, 1945; and Miguel Izard. *Series Estadísticas para la historia de Venezuela, 1970*; as well as Manuel Landaeta Rosales. *Gran Recopilación geográfica, estadística, y histórica de Venezuela*, 1963, contain the principal statistical series of production, trade, prices and other data of the nineteenth and early twentieth centuries.

16. Manuel Vicente Magallanes. *Op. cit.*, 2nd Vol., chapters X-XIII.

17. Ramón J. Velásquez. *La caída del liberalismo amarillo*, 1973. The statistical data are from Veloz, *Op. cit.*; and Izard, *Op. cit.*

18. Domingo Alberto Rangel. *Los andinos en el poder*, 1964, is a standard and somewhat colorful account.

19. Aníbal R. Martínez. *Cronología del petróleo venezolano*, 1970, traces the history of state subsoil ownership in Venezuela.

20. Edwin Lieuwin. *Petroleum in Venezuela: A History*, 1955; and Aníbal Martínez. *Op. Cit.*, provide the details of the concession transactions, production and exports.

21. Luis Cordero Velásquez. *Gómez y las fuerzas vivas*, 1971; and Domingo Alberto Rangel. *Gómez, el amo del poder*, 1975, are recent studies of the Gómez period. Even today, however, it is not possible to find an unbiased account.

22. The studies by Lieuwin. *Op. cit.*; by William D. and Amy L. Marsland. *Venezuela through its History*, 1954, 1974; by Domingo Alberto Rangel. *Op. cit.*, and others, are universal in their condemnation of Gómez and his regime.

23. Pedro Esteban Mejía Alarcón. *La industria del petróleo en Venezuela*, 1972; and Luis Vallenilla. *Auge, Declinación y Porvenir del Petróleo Venezolano*, 1973, and many other studies document the early days of petroleum and its rapid devleopment.

Chapter III

Evolution of Democracy

The evolution of democracy is an integral part of the Venezuelan transformation in recent decades. Many of the policies touching the economy stem from the approach to, and achievement of democracy. Starting with a dictatorship which repressed all opposition, Venezuela developed by degrees a workable party-based political system capable of adjustment and improvement and which has built-in methods for resolving conflicts. The system provided the setting within which the government formulated, promulgated, and implemented efficacious petroleum and economic policies. Evidence of the stability of democracy since 1958 consists of four free elections, three constitutional turnovers of power, two transfers of power to the opposition party, and the current absence of efforts to subvert the system.

Ironic as it may seem, even the 27-year dictatorship of General Juan Vicente Gómez played a role in the development of democracy. Although a regional caudillo himself, he destroyed the system of regional rivalry of the nineteenth century by creating a strong central government and national army. By eliminating all political opposition, he uprooted the century-old parties and conflicts and focussed attention on the disservice of personal dictatorships to all interest groups except the dictator's favorites. A more lenient and less self-serving dictator would have provided a less urgent motive for seeking an alternative political system.

Latent opposition persisted throughout the Gómez regime. An efficient spy system and military might squashed sporadic uprisings and filled the jails with political prisoners. In 1928 university students spoke openly against the regime and gathered political support, but not enough to mount a revolution. In jail and in exile, however, student leaders continued to conspire, formulated a new political system, and formed the nucleus of the group which was to lead the country to democracy in the years to come.[1] So awesome, however, was the dictator's machine that only his death in 1935 initiated political change.

Political Chronology

Military takeovers, unconstitutional procedures, fraudulent elections, denial of civil rights, unwise words and acts, and irresponsible behavior have littered the road to democracy in Venezuela. Scattered among the failures, however, were giant achievements: civil liberties, freedom of the press, formation of political parties, popular election of the president and congress, statesman-like compromises, effective coalitions, and the development of a method for resolving conflicts. The table below and the following paragraphs outline the major events and personalities in the attainment of democracy.

Date	*Chief Executive*	*Type of Regime*	*Method of Succession*
1935–1941	Eleazar López Contreras	Liberalizing Military	Death of Gómez
1941–1945	Isaías Medina Angarita	Liberalizing Military	Election by Congress
1945–1947	Rómulo Betancourt	Civilian-Military Junta	Military Takeover
1948	Rómulo Gallegos	Civilian	Popular Elections
1948–1952	Military Junta	Military	Military Takeover
1952–1958	Marcos Pérez Jiménez	Military Dictatorship	Fraudulent Elections
1958	Military Junta	Military	Military Takeover
1959–1963	Rómulo Betancourt	Civilian	Popular Elections
1964–1968	Raúl Leoni	Civilian	Popular Elections
1969–1973	Rafael Caldera	Civilian	Popular Elections
1974–	Carlos Andrés Pérez	Civilian	Popular Elections

General Eleazar López Contreras was a transition president, occupying the period between the absolutism of Gómez and the near-democracy of General Isaías Medina Angarita. He assumed office on the death of Gómez, and the Gómez congress elected him President in early 1936. Coming to power as the Gómez War Minister, he yet dismantled the Gómez regime. Conciliating the democratic opposition, he yet rebuffed the organization of political parties, stalled on political and electoral reforms, and exiled the opposition. López Contreras walked a tightrope, recognizing and accepting the need to move toward democracy but resisting it in the interest of stability and order, and under pressure from privileged military and economic groups. Clandestinely, however, political groups organized, learned how to operate, and gathered popular support.[2]

General Medina Angarita became president in 1941 upon nomination by his predecessor and ratification by congress. Although he, like López Contreras, represented the Andean military—the heritage of Gómez—his regime brought genuine liberalization. He recognized political parties, including Acción Democrática (AD), which was to become the fountainhead of nearly all parties. Medina emptied the jails, recalled the exiles, eliminated press censorship, and introduced reforms in agriculture, social security and education. In 1942 he promulgated the progressive income tax and his hydrocarbon law of 1943 became the framework for future petroleum policy. The political opposition spoke without fear of reprisal. Under Medina, Venezuela approached democracy. Power, however, remained with the military.[3]

Acción Democrática was the major political party in opposition to Medina, and by 1945 was eager for power. After the collapse of a plan for cooperation between AD and the Medina government in the selection of Medina's successor, AD joined with disaffected young military officers in ousting Medina before his term ended. AD justified itself by arguing that Medina's successor would delay democracy indefinitely.[4] Despite military participation in the takeover, civilians dominated the junta and Rómulo Betancourt of AD was the interim president.

Political and economic reforms came tumbling out of Miraflores Palace. The new government wanted to change everything at once. The junta called for popular elections of the president and congress. New parties sprang up. Labor unions organized. In the first free elections in Venezuelan history, Rómulo Gallegos, novelist and educator and head of the AD ticket, was elected President.

The reforms and political bickering threatened the military and economic interests. The AD mandate spread fear of the loss of power and privilege among important sectors of society. In 1948 the military reclaimed the government, retained power, and governed through a triumvirate. Political leaders again went into hiding, exile, or to jail; political parties went underground; the regime repressed the budding democracy. After fraudulent elections in 1952, General Marcos Pérez Jiménez, a member of the triumvirate, declared himself President and governed with dictatorial powers. His government wiped out the vestiges of democracy, eliminated most of the reforms, and was a painful reminder of the days of Gómez.[5]

The abuses of the dictator offended not only the military but also every other sector of Venezuelan society, including economic interests, the church, labor and the underground parties. During the general uprising in early 1958 the dictator fled the country and the succeeding military junta promised free elections.[6] The parties came out of hiding and conducted a vigorous campaign. They agreed to abide by the election results and the winner agreed to form a coalition government. AD won and Rómulo Betancourt became president in 1959.

The Betancourt presidency picked up where the 1945–1948 experiment left off, but with a significant difference: greater caution in reforms and more sensitivity to vested interests. The previous experiment had taught AD and Betancourt that a successful democracy must serve all interests in society, not just its followers. Nonetheless, in his presidency, the military threatened on the right, the terrorists attacked on the left, and the economy staggered and stumbled. But by compromise, cajolery and cunning, Betancourt endured.[7] In 1964, for the first time in Venezuela, a popularly elected president draped the yellow, blue and red presidental sash over the shoulder of his popularly elected successor.

President Raúl Leoni of AD continued the policies of Betancourt and introduced new reforms. Congress increased its participation in political decision-making. Violence subsided but political controversy was fierce. The party system, however, held together and the political parties matured. In 1969 Venezuela registered another first: Rafael Caldera of the principal opposition party, the Comité de Organización Política Electoral Independiente (COPEI), won the election and became president.

The successful change of party controlling the government helped to consolidate democracy. The new government continued most of the economic and social policies of the previous regimes. In the COPEI government violence disappeared, the professional military accepted its role, and the party system enlarged its effectiveness in resolving conflicts. Democracy matured and AD, which controlled the congress, and COPEI emerged as the dominant political parties.

In 1973 Venezuelans went to the polls for the fourth time since the dictatorship ended and elected Carlos Andrés Pérez as president, as well as an AD-controlled congress. Miraflores Palace witnessed the third transfer of power and the second interparty transfer. COPEI and AD had polled 85 percent of the vote despite a dozen other presidential candidates. In a decade and a half Venezuelan politics had changed from an uncertain experiment in representative government to a robust two-party democracy with accepted norms of conduct.[8]

Political Parties

Two parties, AD and COPEI, dominate Venezuelan political life, although congress has representation from the smaller parties, which also nominate presidential candidates and influence public policy. AD and

COPEI are coalitions, each representing many sectors of Venezuelan society. These differing and sometimes conflicting interests blunt these parties' ideological sharp edges and maintain both of them near the center of the political spectrum. Other parties with a more narrow clientele have more dogmatic positions, mainly left of center. The principal uniting elements of all parties are maintenance of democracy, nationalism, and the advocacy of economic development.

Until 1941 no organized political party operated openly in Venezuela. Political groups blossomed after the death of Gómez, centered around the student movement which had erupted in 1928, but López Contreras denied them legitimacy. The new political forces, however, thrived underground on the regime's liberality and the acceptance of public opposition. The outlook of political leaders opposing the regime was reformist and revolutionary, heavily influenced by leftist thinking. They urged direct elections, more nationalistic economic policies, rapid economic development, and the elimination of Gómez influences in the government and in the economy.

In the election of 1941 Rómulo Gallegos of the major opposition group obtained permission to campaign for the presidency, but the congressional election of the president predetermined the succession of the candidate of López Contreras. Medina Angarita, however, only three months in office, authorized the formation of Acción Democrática, which immediately began to organize a broadly-based national political party. "Not a single district nor single municipality without a party organization" was the AD motto as it mobilized all elements of opposition to the government.[9]

The organization work continued for years, and in the absence of any other major political parties AD had a clear field to penetrate deeply into Venezuelan society. AD leaders went everywhere and talked to everyone, preaching electoral and economic reform. Revolution was often on their lips but the party, at this point, had decided to achieve power legitimately. AD picked up in its net most of the leftist element not already committed to the small Communist party, the majority of peasants, the less ideologically-oriented middle class, and some commercial and private sector support. The center of gravity of the party in these early years was decisively on the left, and it favored higher taxes and more control of the petroleum industry as well as government intervention in the economy to promote growth and to redistribute income.

In 1945 AD opportunistically took advantage of the succession crisis and came to power while organization work was still going on. The government had not effected electoral reforms and Medina was preparing to nominate his successor to congress, tantamount to election. The military, however, split; one element favored Medina but another element rejected the old Andean hegemony. At the instance of these young officers, AD (providing popular and labor support) and part of the military establishment toppled the government. AD commanded four of the seven junta positions as well as the presidency and Rómulo Betancourt became the interim president. The junta changed the electoral system, provided for direct presidential and congressional elections, recognized political parties, and called for elections.

The organization work of AD paid off in 1947 when the Venezuelan people elected Rólumo Gallegos president and AD won control of congress. The popular mandate, however, went to the party's head and the reforms accelerated. The private economic sectors were uneasy about the sweeping economic reforms and the leftist leanings of AD. The church moved into opposition because of AD's anticlerical tendencies. After the elections, the AD hostility toward the military became more apparent. In 1948 the military demanded control of the government and rather than risk civil war, AD succumbed and its leaders went into exile or underground.[10]

For nine years AD operated as a clandestine party and despite repression maintained its organization and the support of the people. It reemerged in 1957 to spearhead the opposition to the dictator in his waning days and supported the interim military junta which promised free elections. During 1958 AD reconstructed the party and conducted a forceful national campaign. In an effort to avoid the problems of 1945–1948, AD, along with other parties, agreed to accept the election results and to form a coalition government if it won. AD captured 49 percent of the presidential vote and also won control of Congress. As agreed, it formed a coalition government.

AD had begun as an umbrella party and at first included nearly all of the opposition to the military government. Later, with the establishment of new parties, and still later, with the responsibilities of government, AD hegemony crumbled. Each new party subtracted from its numerical strength by appealing to some special interest or the inclination of a leader. First, the Communists formed their own party. In the 1945–1948 period, the Unión Republicana Democrática (URD) became the vehicle of Jóvito Villalba, a former AD leader; but aside from his leadership, URD was indistinguishable from AD. COPEI, a Social Christian party, drew conservative, Catholic and private sector support. In the 1958 elections three parties shared 95 percent of the votes, AD leading, URD second, and COPEI third.[11]

AD again began to lose support when it came to power in 1959. In 1960 a leftist group, disappointed with AD reforms, formed a new party, taking substantial numerical strength with it. In 1962 a personality conflict split AD once again and diluted AD congressional control. The main party held together, however, and won the 1963 elections, but with a much reduced strength, polling only 33 percent of the popular vote. During the Leoni presidency the slower pace of economic reform again did not satisfy some AD elements, setting the stage for yet another split which occurred just before the 1968 elections. In that year AD lost the elections. With 28 percent of the vote, AD, however, came within 31,000 votes out of 3.6 million cast of winning the presidency.[12]

The leftists were the principal defectors from the AD ranks. They wanted faster and more radical reforms. The result, however, was that AD moved from its pronounced leftist position of 1945–1948 and early in the Betancourt presidency, to a more center position in the later Betancourt years and in the Leoni presidency. The failure of AD to implement more interventionist economic policies in its 1959–1963 mandate was the cause of many of the defections. The withdrawal from the confrontation with the combined oil and economic interests in 1966 caused other defections. The cause of AD reticence was more practical than ideological; it sensed that more radical economic policies might encourage military intervention and the loss of democracy. Better a more moderate policy and more gradual attainment of goals than endanger the system.

After the splits, the pressure for more radical reforms came from outside the party, leaving AD a more homogeneous and more conservative party. The impulse to reform remained, as did economic nationalism, but within the framework of the existing distribution of power in Venezuelan society. Having failed in 1945–1948 to retain power, even with popular support, AD decided that it must create and maintain conditions in which all segments of society felt that popular democracy conferred more benefits or inflicted less harm than did military government. Its reforms must not alienate important sectors of society. To effect reforms it must govern; to govern it must exercise restraint, and conciliate. AD did not permit itself the luxury of alienating the business community, attacking the military, offending the church, or threatening unduly those with wealth and high incomes.

Economic policies were incremental reforms, changing and improving the existing arrangements at the margin, rather than attacking the citadels of power. AD accepted the private sector and private sector growth, even encouraging it, but at the same time initiated programs, created institutions, and installed controls that vested the public sector with increasing authority over the economy. In the Leoni presidency, for example, AD initially supported a series of policies—tax reform, increased taxes, entry into the Latin American Free Trade Association—

which coalesced domestic private sector opposition, as well as that of the oil companies. Rather than risk undermining the political system, AD backed off.[13]

Even though the petroleum sector had no Venezuelan constituency and was an inviting target, especially in view of the nationalism which permeated all parties, AD moved with caution. The long-run goal of complete control of the petroleum industry motivated gentle and steady pressure, rather than brusk moves which might generate a confrontation. AD recognized the overwhelming importance of petroleum to the economy; significant production cutbacks or a cessation of production would drastically reduce public revenues and create a crisis which might destroy democracy.[14]

Even in political defeat AD acted to preserve the delicate balance in Venezuelan society. In the 1968 elections its loss was so small that a recount might well have indicated that AD was the winner. But the challenge might also have undone Venezuelan democracy, if important segments of society became convinced that AD was unwilling to relinquish power. Better to suffer defeat and relinquish power for five years than to jeopardize democracy and be banished indefinitely.

In opposition, but in control of congress, AD continued to demonstrate political leadership after the 1968 elections. From the AD banks of congress national policy emanated, forcing the executive to acquiesce or to oppose popular measures. Increased petroleum taxes and unilateral government reference price fixing were AD initiatives. The COPEI president almost lost control of the policies of his own government. Even though deserted by many members, including some of its most able leadership, AD in opposition demonstrated the will to power essential to any political party. The elections of 1973, in which AD obtained 49 percent of the vote and elected an AD president and congress, justified its acceptance of defeat in 1968 and its behavior in opposition.[15]

COPEI

The other major party, COPEI, arose in 1946 under the leadership of Rafael Caldera, who, like Betancourt and Leoni, had participated in the student revolt of 1928. He campaigned in 1947 against Gallegos, polling 22 percent of the vote. The party has its roots in the Social Christian movement, a political force in many countries. In Venezuela its greatest strength has been in the more conservative Andean states, the middle and professional classes, and among Catholics, although it is not a church party. Like AD it includes diverse ideological sentiments, focussing on a center position, with fewer leftist leanings than AD.

COPEI entered the coalition government of Betancourt in 1959 after elections in which it won only 15 percent of the vote, about one-half that of URD, the second runner. It earned a reputation for being able to provide able administrators and professionals. In Congress it had only six out of 51 senators and 19 out of 133 delegates. Before the 1963 elections it withdrew from the coalition and carried on a strenuous campaign, again with Caldera. Its share of the vote expanded in 1963 to 20 percent, ahead of URD. COPEI attached such conditions to its continued participation in the coalition that AD sought support elsewhere and COPEI became the major opposition party with greatly increased strength in Congress.

COPEI's internal solidarity contrasts sharply with that of AD. In each election except 1973, COPEI fielded the same candidate and in 1968 he squeaked in, profiting by dissension in AD. Less than a percentage point separated the 29 percent obtained by COPEI from AD. In office the COPEI fame as an administrator dimmed somewhat and AD often maneuvered COPEI into a defensive position. During its administration, however, COPEI, sensing that its conservative image damaged its popular support, turned left and much more nationalistic. COPEI as a party, established its viability in 1973 when, without its illustrious leader, it still polled 36 percent of the vote, more than it had in victory five years earlier.[16]

While these two parties dominate Venezuelan politics there are a large number of other parties. The bones of defunct political parties, coalitions, and fronts litter the history of the past three decades. Some parties have been durable, such as Villalba's URD, which was the number two party in 1947 and 1958. It persists today, still under his leadership and still without ideological or political identifying features. The Communist party has also endured, although the government outlawed it for most of the 1960s.[17]

In other cases a party sprang into existence for a single election or surrounding a prominent personality; electoral failure then killed the party. Even the former dictator, General Pérez Jiménez, organized a party from Madrid, sought office, and the party won some congressional seats. In still other cases, electoral campaigns have brought momentary coalitions or fronts into existence. These combinations of small parties unite behind a single candidate. The parties retain their identity for congressional elections and the front usually dissipates after the election.

In 1947 four parties (AD, URD, COPEI and the PCV) contested in the elections. These same parties fielded candidates in 1958, and in addition three tiny parties won a few votes. In 1963, of the seven candidates for the presidency, four obtained 88 percent of the vote. In 1968 eighteen parties put forward six candidates; four obtained 99 percent of the vote, and the top three, 80 percent. The tendency was toward multiple parties, coalitions behind a few candidates, and concentration of voter preference among even fewer. The 1973 election confirmed this trend when AD and COPEI obtained 85 percent of the vote and 12 parties and their candidates shared the remainder.

The proliferation of parties results primarily from the disorganization of the left. Denied strong influence within the two major parties, bereft often of competent leadership, and in the absence of any unified doctrine, the left carelessly fritters its strength away in personalities and ideological divisions. Bit by bit the major parties espouse and enact parts of the leftist programs, often leaving the left in untenable doctrinaire positions or as shrill and ineffectual critics. Abundant public spending, the growth of the economy, and economic reforms have provided leftist parties with few positions which command any significant popular support.

Organized political activity on the right is limited. A small party has presented presidential candidates, but he is consistently among the least at the polls. Pérez Jiménez' party enjoyed only a brief flurry but found no broad public support. Rightest sentiment is more widespread in the population than party activity indicates, however, because those who hold these views estimate, probably correctly, that their greatest influence is within the major parties rather than in a separate and isolated political party.[18]

Party discipline has been an essential ingredient in the development of parties. When the party has decided on a policy position in its internal councils, all party members support it, on pain of expulsion. No delegate or senator or party spokesmen would think of voting against the party or speak against it. Party discipline enables the government to undertake decisive policies when it controls the executive and the majority in congress, or can garner the support of other parties. The party can confront crises in a unified way and count on the backing of the membership. The party is more important than the personal views of a member.

Party leaders have great flexibility in formulating the position of the party. The chief executive may take strong positions, modify the party position, form coalitions, add or drop policies, and operate with substantial independence and rely on the inner council of the party and the entire party to support him. In this way, the party leader, but no one else, can effectively lead the party, establish its position, and carry out its program with the full backing of the party.[19]

The development of broadly based national political parties, the two largest representing coalitions of interests but with decisive and accepted leadership, is the vital ingredient of the success of Venezuelan democracy. AD and COPEI, along with other ideologically or personality oriented parties, have been sufficient to absorb the complete political interest and activity of the nation. Aside from the abortive military and terrorist efforts to unseat the government in the early 1960s, those who have aspired to political power and influence since 1958 have had to work through political parties. Those outside the party framework effectively excluded themselves from the opportunity to decide policy issues.

Special Interest Groups

While political parties have carried the burden of articulating the interests of the people, various groups having interests to defend or foment and self-appointed spokesmen for the people have also swayed political processes. These groups primarily function within parties or as adjuncts to parties. In some cases they are outside the party framework, but when they endeavor to influence public policy their efforts are directed at political parties. Functioning mainly through formal associations, some groups exercised a decisive influence on political and economic events until 1958. Subsequently, their direct impact has been muted, and while they continue to be an integral part of political processes, they accomplish their mission through the parties. The principal groups are the military, organized labor, the private economic community, professional associations, campesinos, students, the press, the church and government employees.

The military ruled Venezuela without political parties until 1941, and for four years thereafter, with only the beginnings of political organizations. In 1945, however, one faction of the military dared not attempt to take control from another faction without the support of the principal political movement, AD. Against an immature civilian government and in the context of what they considered a direct threat, the military was later able to act without political support and it ruled by force for nearly a decade. Since 1959 the military has had an unused veto over the governments of Venezuela and the actions and policies of these governments.

Conventional wisdom says that the military can take over Venezuela any time it wishes. In the sense of the availability of physical force this may be true; in a political sense it is true only with a high probability of violence, civil war and protracted unrest. In recent democratic governments military interests have been carefully protected. In defense and internal security its voice is decisive and in economic and social policy, it exercises a restraining influence.

Increasing professionalism, dimunition of violence, generous treatment and respect by the government—ten percent of an expanding budget—and the absence of strongly disaffected segments within society with which it might coalesce, have kept the military in the barracks. The success of democratic governments have convinced most military men that the present political system serves the country and the military itself better than intervention and military government. Past failures in military management of government are another motive for letting civilians run the government.[20] Increasingly, military men occupy positions of high responsibility in the administration of the public sector. General Rafael Alfonso Ravard, for example, is the chief executive of Petróled de Venezuela, the new holding company for the nationalized petroleum industry.

Organized labor did not exist under the Gómez dictatorship, began slowly in the two succeeding military regimes, and blossomed into a powerful political force in the 1945–1948 period after backing the AD-military takeover. The subsequent military regime and dictatorship stamped out the movement from 1948 to 1958, although underground labor groups were instrumental in unseating Pérez Jiménez in 1958. Since 1958, organized labor has spread throughout the labor force, nearly one-half of which is now unionized.

Unions exercise their influence within political parties. AD has attracted most of the labor support, although in recent years COPEI has assiduously solicited labor organizations with some success. The smaller leftist parties have enlisted from the ranks of some unions. Organized labor directs much of its efforts at achieving its goals through political action, rather than as a bargaining agent for its members. While contracts and collective bargaining are the norm, labor unrest and strikes are often political rather than economic. Inextricably bound to party politics, organized labor is one of the principal bulwarks of the present political system.[21]

The private economic sectors have two channels of influence into party politics and the government. Some of the giant enterprises, such as the Mendoza, Vollmer and Zuloaga conglomerates, have direct connections with political leaders and strong links with government bureaucracies. Many of these interests date back to Gómez days, and their representatives have played an active role in all governments, military and democratic. Many politicians sometimes attribute a sinister influence to the economic oligarchy while at the same time they rely on it for party funds, to staff important government positions, and for support of the system. For its part, the oligarchy supports democracy partly out of conviction and partly on grounds that it is less of a threat and of greater benefit than any alternative system.

In addition, private economic interests have combined into a powerful organized voice to influence governments. The Federación of Cámaras de Comercio e Industria (FEDECAMARAS) assembled all the independent trade, industrial and commercial associations in 1944. Annual meetings, study groups, publications and statements by spokesmen carry the message of the private sector to political parties, the government, congress and the people. It believes that a strong and growing private economic sector is necessary and that the government, on occasion, has unjustifiably moved in the direction of greater socialization and control of the private sector. Even while maintaining this defensive position, FEDECAMARAS also works with the parties, congress and the government in economic policies, legislation, and in planning. Pro-Venezuela is another business organization which speaks for the private sector.[22]

Students, especially university students, were the vanguard of opposition in the Gómez era and formed the leadership nucleus which seized power in 1945, lost it, and regained the government in 1959. The Generation of 28 consists of such leaders as Betancourt, Caldera, Villalba, Gonzalo Barrios and many others who have dominated the leadership of nearly all parties up to this time. Students have been the catalyst of unrest and discontent against all governments and a source of ideas on which all parties feed. Most student activities since 1958 have operated under party auspices.

The smaller leftist parties tended to have more appeal for students and even within the major parties, students tend to occupy the left end of the spectrum. Students regard themselves as spokesmen for the people and by taking strong stands on controversial issues, in and out of party councils, they can sway but not necessarily determine policy. Student activities outside the party framework have declined since 1958.[23]

Military regimes censored and controlled the press and other communications media, but from 1945 to 1948, and since 1958, they have operated freely and with few inhibitions imposed by the government. Enterprises, newspapers, magazines, and radio and television stations do not take political or policy stands, but they make their positions known through signed editorials and commentaries. A newspaper favoring AD will employ primarily commentators favorable to that party. Most people do not read newspapers or magazines and most radio and television is entertainment. But as a source of information and ideas, and with nearly complete editorial freedom when they choose to exercise it, the media support democracy.

The government does influence the media through access to information and through the pressure it can apply as the largest single advertiser. A newspaper may restrain its criticism if it knows through informal channels that it may lose its largest customer. On the other hand, all parties and all political leaders have access to the press which reports indiscriminately what these leaders say. But the media as a group, performs an educational rather than an activist political role.

The traditionally weak Venezuelan church has played a limited political role. It did support the military takeover in 1945 and it assisted in ousting Pérez Jiménez later. COPEI captures the vote of many Catholics and it has the support of the church. Professional associations, such as organizations of medical doctors, engineers and economists, take political positions and divide along party lines. Patronage government employees are a potent political force, especially at election time. Most organizations for whatever purpose—faculty bodies, private clubs, fraternal orders—tend to be politicized.

The intensity of political sentiment runs high in Venezuela and finds expression in all income and social classes, occupations, levels of education and organizations. Political matters are topics of conversation anywhere, anytime. People tend to follow party lines, with each party having two groups of supporters: members and sympathizers. The former toe the party line; the latter are less reliable but tend to accept the party position. The policy positions accepted by individuals are generally those espoused by parties, with only a few accepting and rejecting positions of two or more parties on the basis of independent thought.

Government Structure

The formal structure of the Venezuelan government has changed little over the decades. Until 1953 it was the United States of Venezuela and in that year it became the Republic of Venezuela, a change in name only. Officially, it is a federation of 20 states, a federal district, two territories and 72 islands. In fact, it is a highly centralized government with other political subdivisions—states, districts and municipalities—in a secondary position and dependent on the central government, since nearly all taxing power and spending are in the hands of the central government.[24]

Venezuela has a written constitution. The last one was promulgated in 1961 and is the 25th in line since independence. The new constitution aspires to guarantee continued popular democracy, direct elections, checks and balances among the branches of government, division of powers among levels of government, and specifications of responsibilities of elected and appointed officials. The three branches of government are the executive, the congress, and the courts. The last is exclusively national. States have elected legislative bodies, but the national executive appoints state governors. Congressional elections are direct, as are those for the chief executive. There is no vice president.

The constitution lays down the basic legal framework. Congress passes laws and its laws can become effective even over presidential veto. The President can issue decrees which have the force of law and do not require congressional approval. Public law often goes through stages: the constitution has a general legal provision, for instance, for a social security system. Later the congress passes a law elaborating upon the constitutional provision. Still later the President issues decrees which make the law operational and provide for the administrative details. Then, vast amounts of administrative law and detailed regulations issue from the bureaucracy.

The President has great power as head of the government, the armed forces, and the finances of the country. A council of ministers, appointed by him, assists him. The most important ministries are finance, defense, public works, mines and hydrocarbons, and other sectoral ministries. Congress has the power to censure ministers but not remove them. The President has no party position while he is President, but informally exercises great influence in party matters.

Congressional power resides primarily in its capacity to approve, reject, or alter the budget proposed by the President. Congress can also initiate laws having policy implications even if the President and his party disapprove. In addition, many laws previously enacted contain provisions for congressional approval of specific acts of the President or his ministers. Congress consists of two houses: the senate with two senators from each state, and the chamber of deputies with membership based on state population. The congress has committees, including a watchdog committee which works while congress is not in session. Political party factions within congress control congressional deliberations, and voting follows party lines.

In addition to ministries, autonomous institutes and government corporations conduct government business in highly specialized fields. Each ministry is nominally in charge of several such organizations, but in fact they operate independently. There are more than 126 organizations of this type, including those which operate steel mills, petrochemical plants, tourist promotion activities, and regional development plans. These organizations, like the ministries, are subject to budgetary control within the executive department as well as oversight by the congress.[25]

A policy measure, such as new tax rates, a new institution, or a government project normally originates with the president. The President and his council of ministers, with the assistance of professionals and technicians from one or more ministries consider the measure, modify it, and finally approve it. The approval takes the form of a draft decree, which, after further consideration inside the executive department and consultation with congress, becomes law with the signature of the President and appropriate ministers. In other cases, the approval takes the form of legislation which is proposed in congress by the President's party. After consideration by committees and debate in congress, in the press, and among the interests affected, it is passed and when countersigned by the executive department becomes law.

Resolution of Conflicts

Before 1935 the method for resolving political and policy conflicts was force and terror. Between 1936 and 1945 military presidents employed conciliation backed by force, and in the democratic interlude of 1945–1948, AD and the military, enjoying great popularity, rode rough-shod over opposition. Then again from 1948 to 1958 the military government employed violence and force. Subsequent governments have developed rules of the game to resolve controversies. These rules, accepted by political parties, interest groups, and the electorate are the key to the viability of Venezuelan democracy.

López Contreras inherited the military strength of the previous regime, and its use (or threatened use) was the final arbiter. He did not inherit the Gómez espionage apparatus, however, which was so effective before 1936 in eliminating problems and people before conflict arose. Since military might was such a blunt instrument, López Contreras relied heavily upon his own political acumen, personal capability, and the prestige of his office to conciliate contending interests. Behind him always, however, was the military.

No institutionalized form of conflict resolution developed. In the central bank controversy, for example, in the late 1930s, the antagonistic forces spent themselves in public controversy and when all the arguments were in, López Contreras decided to establish the bank. The same thing happened in the petroleum law of 1938. Nationalistic political elements debated labor unions, petroleum policy, social security, agricultural reform, economic policies, but in the end the president accepted some, rejected others. Unprepared to face awesome military might, the losing side accepted the verdict.

Authority continued to be the means for resolving conflicts under Medina Angarita. The income tax and the 1943 hydrocarbon laws did not emerge from a controversy in which opposing forces reconciled their differences. They were measures desired by the chief executive and a compliant congress acquiesced in his request. Emerging political forces opposed the hydrocarbon law and voted against it after failing to effect any changes in its provisions. Medina was, however, more attentive to the opposition voices than was his predecessor.

In both regimes the weight of the presidency, backed by military forces, was the decisive element. In both periods, however, the rising political temperature and the desire of the presidents to conciliate the political opposition rather than eliminate it limited the choices of the government. Many conflicts did not arise because these presidents deliberately restrained themselves. These self-imposed constraints by military presidents who could have behaved in a much more arbitrary manner, set the stage for the evolution of rules of conflict avoidance and resolution which developed later within the context of political parties.

The AD-dominated junta in 1945, and the AD government in 1948, moved far and fast, rolling over opposition in their frenzy of reform. Once again, the weight of the executive, backed by the military (until it turned on the regime) resolved conflicts. The conviction of AD that it had all of the important segments of society behind its government led, however, to intemperance, with little or no effort to hear differences, much less resolve them. This failure to develop conciliation techniques to protect various economic groups, the church, and the military, was the principal cause of the 1948 military takeover.

The 1948 military junta reverted to the methods employed by the 1936–1945 military regimes, but with less success. The dictator, beginning in 1952, turned increasingly to the Gómez model of cronyism and repression. Pérez Jiménez ignored most interest groups, banished political parties, and made no effort to pacify those opposed to the regime or its measures. For a time, the universal solvent of money stalled off the inevitable. It was in this period, however, that the major parties, operating clandestinely, developed an informal set of rules which would regulate their behavior when they were permitted to function openly again.

The rules of the game are uncodified principles which the major political parties have accepted and which form the basis for the preservation of democracy. The first rule is that the maintenance of political democracy is more important than any party, personality, or policy. A party accepts defeat and the position of a member of a coalition, or open yet loyal opposition, rather than endanger the system. And no person is above party or above the system. Additionally, no policy favored or opposed by a party is worth bringing down the system for. The parties recognize that democracy is still a fragile institution in Venezuela and that no party has the right to threaten it in order to achieve or maintain power or to insist on a specific policy.

Another rule specifies that political activity must be confined to political parties. Street mobs, student violence, calls to the military, and the threat or use of force are inappropriate methods for resolving conflict. Contending political parties carry conflicting policies to the people, and voters decide, for a specified period of time, which policies shall be pursued.

Winners and losers continue to disagree, the winner having the right to promulgate its policies and the loser to attack the policies and propound its own. A popular mandate does not silence the opposition. And vital interests of all segments of society must be taken into account by the winner. In addition, the leadership of the parties have special privileges in formulating and changing programs and policy proposals for the party. When the leadership has decided on the position of the party, the membership accepts it.

Abundant and growing public spending, and high and increasing income levels have dulled the sharp edges of many conflicts and avoided others. Governments have employed their financial resources to buy off the opposition and to reduce the potential damage to interest groups of its policies. In land reform, for example, the government buys private land. Government orders keep the private sector humming, church allowances are more than adequate, military pay and prerequisites are generous. The government can afford to implement programs important to the opposition and employ armies of people. More stringent economic conditions would have generated more conflict and discontent and rendered their resolution more difficult.[26]

Summary

More than a century of dictators did not prepare Venezuela well for political democracy but it did demonstrate the high cost of repressive governments. When popular political sentiments began to emerge in 1936 they found expression in the formation of political parties. Two military regimes accepted increasing public opposition and initiated the establishment of methods for hearing and conciliating conflicting policy views.

The short democratic experiment installed by a military takeover prematurely conducted a reform program with insufficient guarantees to those threatened by the rapid changes. AD misinterpreted a popular mandate to mean the right to ignore and demand acquiescence from the opposition. Another hard lesson in dictatorship taught AD and other major political parties the necessity of rules which placed the integrity of popular democracy as the most important policy.

Political parties are the vehicle through which Venezuela has attained and so far maintained democracy. All other interest groups have accepted the predominant role of the parties and the parties have accepted, and indeed insisted upon, the subordinate role of interest groups, while simultaneously soliciting their help and serving them. The electorate and the politically influential parts of Venezuelan society regard democracy as a highly imperfect but workable system which inflicts fewer losses and confers greater benefits for all participants than other political forms. Economic progress has played a central role in making the system function and its rules effective.

NOTES

1. John D. Martz. *Acción Democrática: Evolution of a Modern Political Party in Venezuela*, 1966, tells the story of early political development; as does one of its principal leaders, Rómulo Betancourt in his *Venezuela: política y petróleo*, 1969.
2. Martz. *Op. cit.*, pp. 17-41.
3. Daniel H. Levine. *Conflict and Political Change in Venezuela*, 1973; David Eugene Blank. *Politics in Venezuela*, 1973; and Leo B. Lott. *Venezuela and Paraguay: Political Modernity and Tradition in Conflict*, 1972, all treat the foundation years of political parties.
4. Rómulo Betancourt. *Op. cit.*, pp. 232–234.
5. Winfield J. Burggraaff. *The Venezuelan Armed Forces in Politics, 1935–1959*, 1972, Chapter V.
6. Winfield J. Burggraaff. *Op. cit.*, Chapter VII.
7. John D. Martz. *Op. cit.*, Chapter VIII; and Rómulo Betancourt. *La Revolución Democrática en Venezuela*, 1969, in four volumes, tells the incredible story of this government.
8. Daniel H. Levine. *Op. cit.*, Chapter IX.
9. Cited in John D. Martz. *Op. cit.*, p. 49.
10. John D. Martz. *Op. cit.*, Chapter III.
11. John D. Martz. *Op. cit.*, Appendix, Table 3, p. 369.
12. David J. Myers. *Democratic Campaigning in Venezuela: Caldera's Victory*, 1973, Appendix, Table 27.
13. David J. Myers. *Op. cit.*, 149–150.
14. Franklin Tugwell. *The Politics of Oil in Venezuela*, 1975, pp. 44–47.
15. Some of the observations are the result of the author's annual visits to Venezuela since 1962, to the reading of the Venezuelan press, and to conversations with Venezuelans in many walks of life.
16. Press reports, December 1973.
17. David Eugene Blank. *Op. cit.*, pp. 116–120.
18. Conversations with Venezuelan political leaders, 1962–1975.
19. David H. Levine. *Op. cit.*, Chapter III.
20. Winfield J. Burggraaff. *Op. cit.*, pp. 190–205.

21. Thomas E. Weil. *Area Handbook for Venezuela*, 1971, pp. 363–384.
22. Thomas E. Weil. *Op. cit.*, pp. 384–385.
23. David H. Levine. *Op. cit.*, Chapter VI and VII.
24. Thomas E. Weil. *Op. cit.*, pp. 405–415.
25. Thomas E. Weil. *Op. cit.*, pp. 205–230.
26. Franklin Tugwell. *Op. cit.*, pp. 164–165.

Chapter IV

Petroleum Policy

Beginning in 1936 government petroleum policy was an experimental and evolutionary process aimed at increasing public revenue and vesting greater control in the government, culminating in nationalization in 1975. This policy of unrelenting pressure on the petroleum companies characterizes the periods of 1936 to 1948, and 1958 to the present. During the military regime (1948 to 1958) the government relaxed petroleum policy. Except in that decade, the government systematically usurped the companies' decision-making powers and forced them to yield a greater share of profits without disrupting the industry.

The Policy of General Gómez

When General Gómez took power in 1908 no petroleum industry existed in Venezuela. The Compañía Petrolia del Táchira operated a few wells and a tiny refinery under an 1878 concession. The government had granted 1,312 concessions between 1878 and 1920 when the first petroleum law replaced the mining laws that governed all subsurface activities. Foreigners drilled their first successful well in 1913. Drilling intensified and discoveries of new fields came every year. A trickle of exports began in 1917. The British and Dutch were the first concessionaires; the North Americans muscled in later.

Production and exports, as well as royalties, climbed. The treasury bulged and General Gómez began paying off foreign debt. He also made money on the side in the concession business. Exports during his regime totalled 1,148 million barrels and the government collected Bs476 million, or Bs0.41 per barrel. Royalties were from 8 to 15 percent, depending on the concession, and the companies also paid surface and exploration taxes and other minor fees.[1]

The petroleum policy of General Gómez was: to leave the petroleum business to the companies, to make money from concessions and production, and to encourage the entry of foreign capital. The legal position of the concessionaires and the tax and royalty obligations of the companies were confusing. The government had granted concession under many mining laws and under petroleum laws of 1920, 1921, 1922, 1923, 1925 and 1935. Concessionaires had bought, sold and traded concessions, sometimes under questionable circumstances, and the legality of many concessions was in doubt. Concessions were for different periods of time and called for different royalties and taxes.

The administration of petroleum policy served the companies with one notable exception. Dr. Gumersindo Torres was a nationalist who believed that Venezuela should benefit more from its oil. In 1917, as Minister of Development, he announced a new policy, and in 1918 he issued new regulations, including concessions for only thirty years, surface taxes that varied with the rate of production (a forerunner of the income tax), royalties that varied on distance from wellhead to terminal, and cancellation of concessions not developed within three years. He even included conservation measures in the regulation.

In 1920 Dr. Torres continued his reform in the first petroleum law. It increased surface taxes, permitted surface owners to obtain concessions on their own land, reduced the size of concessions and increased the size of national reserves, and reduced the list of duty-free imports for the companies. The oil companies objected, took their case to the national legations, and pressured General Gómez. He fired Dr. Torres and asked the oil companies to write the 1921 law which was much more favorable to the companies. They also edited the 1923 law which became the basis of petroleum law for two decades.[2]

Through the 1920s, the foundation period of the industry in Venezuela, the companies had their way; but in 1929 Dr. Torres was back as Minister of Development. He increased taxes, established a technical petroleum office in his ministry, and began field inspections and supervision of company facilities. Again the companies protested but Dr. Torres would not back down, although the great depression undermined the effectiveness of his measures.

The great sweep of petroleum policy from 1908 and 1935 was exploitation, corruption, conspiracy between General Gómez and the oil companies, and favorable treatment of the oil companies. Dr. Torres, for all his valiant efforts, was a small candle in an otherwise dark night. Edwin

Lieuwin in his *Petroleum in Venezuela* and Ròmulo Betancourt in his *Venezuela: Política y Petróleo*, and many others document the chicanery and dishonesty, both of officials and the companies.[3]

In addition to the precursor role of Dr. Torres, two aspects of the Gómez period bear on later policy. General Gómez modernized the government. He established national institutions and centralized decision-making in the ministries. He reformed tax collecting and the administration of public funds. In addition, the Gómez petroleum policies induced the companies to come to Venezuela and make large investments. His modest royalties and taxes, generous concessions, and permissiveness in the operation of the industry made Venezuela the largest exporter and second largest producer of petroleum in the world by 1935.

The Great Debate

In the more liberal atmosphere of the López Contreras regime a great debate on petroleum policy began. The petroleum novel had been the vehicle for earlier criticism but now with a freer press, the debate came out in the open, in newspapers, magazines, on the street and in parlors, and in government offices. People began to question government petroleum policy, to argue that it was too lenient, that the government did not collect enough taxes. The government modified the petroleum law slightly in 1936, but more important, it began to enforce existing law more meticulously. In addition, the government established a geological technical service and published the first geological map. It began to train its own geologists, technicians and petroleum engineers. A new labor law led to the formation of a petroleum workers' union.

The debate continued in 1937 and the government accelerated its demands on the companies. It began to examine carefully data submitted by the companies, and brought suit against the companies for additional taxes because of incorrectly calculated costs. Congress set up commissions to study the petroleum industry. The press kept up a constant din for a more vigorous petroleum policy. Bit by bit the government became more aggressive, declaring the private oil company roads to be public roads and ordering the companies to clean up rivers and lakes they had fouled.

The 1938 petroleum law doubled royalties and raised taxes. It authorized the government to go into the petroleum business and ordered the construction of a national refinery. The government suspended concession granting and began to require prior permission to initiate drilling, begin construction, or transfer concessions. In 1940 when the president started to implement all the new measures the companies announced that further petroleum development would cease until the government changed the law.[4]

The 1943 Petroleum Law

The 1943 petroleum law broke the deadlock between the government and the companies. Until that year thirteen mining and petroleum laws governed taxes, concessions, royalties and regulations of the industry. Both the government and the companies had an interest in simplifying the system, establishing new rules, and backing off from the confrontation. The 1943 law, while officially a law, was also an agreement between the companies and the government and was the basis for petroleum regulation until nationalization.

The government had specific goals that it wanted to achieve in the law. These included more revenues, an equal participation in petroleum revenues, a simplification of the concession system, and a restatement of the rights of the nation. The companies also had goals. They wanted concessions of longer duration and cleared of legal entanglements. The companies were under pressure from the United States government to make an agreement, since it was wartime and the U.S. wanted a secure source of petroleum. The Medina government wanted an agreement because it needed revenue badly and the political opposition was pressing for strong unilateral action.

The 1943 petroleum law wiped out all previous laws on petroleum and established a new system. The old concessions were eliminated and the law authorized granting new concessions under uniform conditions. As important, however, as the law itself, was the process by which it came into existence. The government and the companies negotiated the agreement and established the basis for future negotiations which lasted until nationalization. The government did not renounce unilateral action but in this law, and at critical junctures thereafter, negotiations led to acceptable solutions without disrupting the industry.

In the bargain the companies obtained much of what they wanted. They received new concessions for forty years, clear of legal taint, and the amount of concession land was greater than before. The new concessions were solid blocks, not the checkerboard alternating blocks as before. The companies could now look forward to uninterrupted expansion.

The benefits to the government were also great. The tangled web of concessions, royalties, taxes and regulations was replaced with a simple uniform system with no economic, fiscal, or legal doubts. Petroleum became a public utility. The government could enter the petroleum business at any time. The law established freedom of exploration and the necessity for concessions in order to produce petroleum, along with total acceptance of risk by the concessionaire.

One of the most important principles established in the law was the principle of reversion. The concessions would last forty years, and at that time the concession and all property associated with it would revert free of encumbrances to the state. Dr. Torres first introduced the idea of reversion in his law of 1920, but that law the oil companies rewrote the following year. But in 1943 it was clear and unequivocal: the companies could extract oil owned by the nation from the nation's property for a specified period. Then everything, including all improvements and company property, belonged to the state. This provision of the 1943 law was the basis of the 1972 reversion law.

Fiscal benefits included an increase in royalties to one-sixth part of production in cash or crude oil, at the option of the government. Surface and exploration taxes also increased. The law gave the President the faculty to exonerate, at his option, petroleum company import duties, replacing automatic exoneration as a part of the concession. Fiscal income per barrel increased 35 percent in 1944. The new fiscal package in the 1943 law, combined with the income tax passed earlier, resulted in a 77 percent increase per barrel between 1942 and 1944.[5]

The Income Tax

The tax system of Venezuela in 1942 was only a minor modification of that which existed in the early days of the republic. Heaviest reliance was on indirect taxes, such as import duties. The sale price, of course, included this tax, making it regressive since the tax impinges more on low than on high income groups. The growth of the petroleum industry had provided additional taxes, including concession, surface, exploration and other minor taxes. The principal government income from petroleum was the royalty, which was not a tax but rather a payment for national patrimony, founded on the royal one-fifth of colonial days.

In 1936 General López Contreras promised tax reform and did not do it. When he took office in 1941, General Medina Angarita also promised tax reform but then began negotiating a foreign loan in order to increase spending without more taxes. The government was fearful of the reaction of the petroleum companies to increased taxes. In 1941 the fiscal income of petroleum was Bs3.37 per cubic meter, about the same as it had been since 1936. The sale price was Bs18.00, meaning that government participation was less than 19 percent. The same year the net income of the petroleum companies was 35 percent of the price, about double that of the government.[6]

Under strong attack by the political opposition, the government accepted the idea of fiscal reform. The result was the first effective income tax among the less developed countries. The government passed the law in June, 1942, to take effect in 1943. Tax rates were low. A part of the tax was invariable; another part varied with reported income and was progressive. Officially the law applied to all natural and juridical Venezuelans. In fact, the government collected taxes only from the petroleum companies.

Several factors favored the introduction of the income tax in Venezuela. First, it was unquestionably an act of sovereignty and within the constitution. The oil companies protested but they had no recourse. Secondly, it was relatively simple. Even though technically a general tax, it was in fact a tax on a small number of enterprises about which the government was already collecting economic and financial information. Thirdly, the oil companies, while never liking any tax, opposed the income tax less than an equivalant increase in royalties.

The oil companies in Venezuela were the property of oil companies in other countries, especially the United States and Great Britain. The parent companies had tax obligations in their home countries. Royalties and some taxes paid in Venezuela, as costs of production, reduced the net income on which the countries paid taxes at home. Income taxes, however, the home countries treated as an offset to taxes at home. If the companies did not pay taxes in Venezuela, they would have paid as much or more at home. Thus, in the consolidated accounts of the parent company, the Venezuelan tax was not an increase in taxes, but rather a payment to a different tax collector.[7]

The Trienio, 1945–1948

In 1945 the political opposition suddenly became the government. Those who had argued so vociferously against the oil companies in years past, such as Rómulo Betancourt and Juan Pablo Pérez Alfonzo, were President and Minister of Development respectively. The oil companies were uneasy, fearing drastic actions from these hotheads. Realizing the critical importance of petroleum to the economy, however, the new government restrained itself. It accepted the reform of 1943, which Betancourt and Pérez Alfonzo had opposed earlier, and did not make any strong new moves against the companies.

The revolutionary government needed revenues desperately to support its reforms in other areas. The petroleum industry was the only source. Two months after taking office the government imposed a special tax on oil company profits, equivalent to two-thirds of the previous year's payments. For the first time the government shared equally in oil profits, one of the aspirations of the 1943 petroleum law.

In 1946 the pro-labor government strengthened the unions in the petroleum sector and encouraged the formation of a powerful federation of petroleum unions. The government coerced the companies into recognition of the unions and the companies and the unions signed a collective bargaining agreement. At the end of the year the government increased income tax rates and began to collect three and one-half times as much per barrel as did the Gómez regime in 1935. Government enforcement of petroleum, tax and labor laws became more strict.

The revolutionary government maintained steady pressure on the companies for conservation and field supervision, and accelerated the training of Venezuelan professionals for the industry and government. Unions and the companies signed another collective bargaining agreement, with increased worker benefits, in 1948. Late in the year the congress approved the resource policy of the Minister of Development. It included a policy of no more concessions, establishment of a state petroleum company, 50-50 participation in oil profits, and government sale of royalty crude. Congress passed a permanent additional tax to assure the minimum 50-50 split. Two weeks later the government fell.

Royalties and Prices

The original idea of the royalty was a division of the proceeds of the mines. If the operator pays in money and not in product, the problem of price arises. This should be a simple matter; the operator sells all of the product on the market and gives the government the same share as the division of product calls for. This simple procedure does not work very well in the case of primary products subject to further processing or in the case of imperfect markets.

The petroleum industry is a vertically integrated industry. The same company that conducts geological surveys and drills exploratory wells, also drills producing wells, extracts crude, transports and stores it, refines it into many products, ships both crude and products, stores them again, refines them further, distributes them, and sells the products at wholesale and retail. The company may have many names, operate in many places and have a complex legal and administrative structure. It is, however, one continuous operation. Not all companies operate this way; some operate on only a part of the chain or even at only a single stage.

The real market is the one where the company sells to a final user, an industrial buyer of lubricants, a consumer of fuel oil, a purchaser of gasoline for his car. The prices of petroleum products registered in these markets reflect backwards to determine the price of crude and petroleum at intermediate stages. Where one company controls the entire chain, the prices that one part of the company charges another, have limited economic meaning. Even in those cases where there is an independent market at some intermediate stage, the price there tends to follow the prices established by the large integrated companies.

The petroleum industry consists of a very few large firms, firms so large they dwarf many of the countries in which they operate. These large companies determine the price structure, both for final products and at intermediate stages. In this kind of market structure the price, within broad limits, is indeterminate. Falling demand or increased supply will drop the price, and rising demand and short supply will drive the price up. But at any given time, the price of petroleum is whatever the companies say it is and the companies set a price which serves their goals.[8]

The principal goals of the companies are profit and growth. Profits are necessary to keep owners happy, to protect the integrity of those who operate the company, and to assure growth. Growth is necessary to provide power and prestige, high salaries, mobility and open vistas for management. The price that serves these interests best is a low price, a price which will encourage the consumption of petroleum products and other products which use petroleum products. The company will therefore deploy all the power at its command to encourage consumption, keep prices low, and control the production process.

In order to keep prices low, it is necessary to keep costs low and to control them. One cost which the companies endeavor to keep low is the cost of the crude oil, that is, the costs of exploration and extraction and the payment to the owner of the crude. For many years the companies in Venezuela used whatever instrument that came to hand to keep royalties, taxes and other payments to the government low. One such instrument was the price of crude itself. If the price at which companies sell crude to themselves for refining is low, then the royalty payment is low and this assists the company in keeping the price of final products low while maintaining profits and growth.

As the owner of the oil, Venezuela's interests conflicted with those of the companies. Venezuela wanted revenues, and the higher the price used in calculating royalties and taxes, the higher the revenues. But the companies set the price. Until 1942 the companies used the New York crude price. In that year the government and companies agreed to use the East and West Texas crudes as the pricing base. Venezuela was still not satisfied and believed that the companies were underpricing crude to avoid payments to the government.

All petroleum laws recognized the right of the government to take its royalty in crude. The companies believed that the country would not know what to do with the crude and that the difficulties of marketing protected their royalty pricing system. Following the second world war the demand for petroleum was strong and the price tended to drift upward. Dr. Juan Pablo Pérez Alfonzo took advantage of this favorable market and announced in mid-1947 that Venezuela would take one-fourth of its 1948–1949 royalty in crude and would offer it for sale directly.

The companies objected strenously but they were in a real dilemma. If they stuck with the East and West Texas formula, the government might take even more royalty crude, go into business, and become troublesome in international markets. If they agreed to pay a higher royalty price, then they were admitting that the old price formula had underpriced the petroleum. Meanwhile the government energetically signed contracts to deliver crude, made barter deals, and noisily disposed of its crude oil. The companies finally capitulated. In negotiations with the companies the government raised the price of crude for royalty payments, some by as much as one-fourth. Venezuela had found an important key to greater participation in petroleum revenues and at least a part of the Achilles' heel of the oil companies.[9]

The Military Regime and Dictatorship, 1948–1958

After the activism of petroleum policy in the democratic experiment, the miltary regime from late 1948 to early 1958 was tranquil. It was under heavy pressure from the oil companies to undo the reforms introduced earlier, particularly the 50-50 profit split, royalty pricing, refining in the country, labor benefits, and the granting of new concessions. They were successful in part.

For reasons not wholly related to the well-being of the petroleum companies, the military regime and dictatorship clamped down on labor unions. As close allies of the now outlawed political parties, unions were a threat to the government and except for government front unions, organized labor officially disbanded but continued to operate underground. The rambunctious labor activity of trienio faded into the past. The petroleum companies benefitted.

The companies pressed the government on concessions; their long-term plans required continuous additions to reserves. So long, however, as fiscal income covered expenditures the government delayed. In 1956 the government's appetite for gargantuan public works outran income and the companies obtained their concessions. In 1957 more concessions went on the auction block. Earlier the regime had taken a backward step in royalty pricing. Complaining that their oil was at a competitive disadvantage, the companies asked for a change in the formula to their advantage; the government accepted it.

The government established the Ministry of Mines and Hydrocarbons to handle petroleum affairs in 1950, signed a new trade bargain with the United States (eliminating petroleum quotas and reducing tariffs by one-half) joined the InterState Oil Compact Commission (an organization of petroleum producers), sent a mission to the Middle East to talk to oil experts there, established the Venezuelan Petrochemical Institute, made minor modifications in tax and petroleum laws, but held to the basic 50-50 split. With expanding exports and rising prices, tension between the companies and the government relaxed.[10]

The Interim Government, 1958

The interim government that ousted the dictator set itself the task of maintaining order and providing the framework for free elections. It made no major moves in petroleum policy at first, although it established commissions to study the industry, talked to the United States about the latter's trade restrictions, and increased income taxes slightly. After a detailed examination of the 1956–1957 concessions, the interim government decided that they were legal.

Once the elections were over it did make a move, a shattering one to the oil companies. In the ten years since Venezuela introduced the 50-50, many other countries had followed its lead. Reluctant at first, the companies accepted and then enshrined the formula. Suddenly the interim government increased the government share to two-thirds. The companies challenged the government and asked for reconsideration; the government refused to reconsider and declared the President of Creole persona non grata. He left, muttering that the companies would defend themselves. It was not a good omen for the beginning of the Betancourt presidency.[11]

Petroleum Strategy

In the great debate following the death of General Gómez, in the democratic experiment of 1945–1948, in long years of hiding and exile, a petroleum strategy emerged. Its purpose was to provide the framework for specific policy measures for dealing with the petroleum industry. The strategy was not the work of any single person, but Rómulo Betancourt, the incoming president in 1959 and Dr. Juan Pablo Pérez Alfonzo, his new Minister of Mines and Hydrocarbons, were leading lights.

The strategy operated within limitations: the government had to respect the petroleum law of 1943 and let the concessions run their course. The government could not nationalize without compensation, and a large and burdensome debt was not acceptable. Furthermore, the government did not believe it had the technical capability to operate the industry and considered even a brief interruption of the industry disastrous. Thus, in the 1960s the government had to tolerate private companies.

Within the framework of honoring all previous agreements the government strategy was to gradually increase the participation of the government in petroleum revenues and shift decision-making in the industry to the government. In some cases the government would take unilateral actions; in others, the government hoped to achieve its goals through negotiations. The government believed that pragmatic and experimental policies would best serve its interests and that a succession of such policies would convert the petroleum industry into a regulated public utility.

The first element of the strategy was the restriction of the size of the petroleum enclave. The government decided to grant no more concessions, a decision subsequently adhered to by every government. Since the companies returned concessions that were not promising, the area under concession would gradually decline. Reserves would also decline and eventually even production, without an alternative to the private companies. Since the concessions would not expire until 1983 and 1984, with some expiring in 1996 and 1997, the government had twenty years or more to find a solution.

One part of the answer was the service contract. Instead of a concession the government would contract with private oil companies to perform specific tasks for which payments and responsibilities would be determined in advance. In the Betancourt presidency service contracts remained a vague policy objective. Congress legislated the general terms of service contracts in 1967 and subsequently the government negotiated several with private companies. Companies, however, were reluctant and the government was uncertain, especially in view of the opposition of Dr.

Pérez Alfonzo, who had originally proposed them but later changed his mind. The service contract remains, however, the method for continued cooperation between the government and the private oil companies.[12]

Government Enterprise

The future of the petroleum sector was, however, in the public sector, since the service contract was only a temporary expedient. The only way to capture the entire difference between price and cost was to place exploration, extraction, refining, transportation and marketing in government hands. In 1960, therefore, the government created the Corporación Venezolana de Petróleo (CVP) as the nucleus of the public petroleum sector. The government granted it concessions and authorized the CVP to engage in the entire range of petroleum activities.

CVP was an operating oil company, in competition with the private oil companies. In that capacity it has grown rapidly; in 1975, while still small, it was the fifth largest in Venezuela. But in addition, CVP served other purposes. The government assigned it the task of overseeing the service contracts, operating an increasing proportion of retail distribution of petroleum products in Venezuela, providing training for specialists, technicians, and professionals in the petroleum industry, and serving as a measuring stick for the private oil companies.

When the government took over the private companies at reversion, CVP was to be the indicated organization to operate the petroleum industry. CVP, however, did not measure up to government expectations, and in 1975 the government formed a new entity, Petróleos de Venezuela (Petroven) as a holding company for the nationalized private companies. CVP continues to operate as a part of Petroven.

Taxes, Prices and Participation

The Venezuelan strategy envisioned the operation of the private petroleum industry as a public utility until the government assumed full responsibility. In this interim period the government wanted to regulate the industry so that profits to the private companies would return about 15 percent on investment, a figure the government estimated was appropriate to a regulated public utility. The remainder of the difference between total petroleum revenues and total costs the government wished to have as public revenue.

The two instruments with which the government expected to capture revenue were taxes and prices. Royalties were a part of concessions and changes were not possible. Taxes, however, could be changed by legislation or executive decrees, and the government and the companies could negotiate on the prices employed in income tax calculations. Beginning in 1948 the government minimum share was 50 percent. The 1958 military junta raised it to two-thirds. During the 1960s the share gradually increased as tax rates went up and reference prices replaced market prices in calculating tax obligations. In the period preceding nationalization, the government share was 85 percent of petroleum revenue.

Market prices of petroleum reached a peak in 1957. In the declines that followed at first the government endeavored to prevent the company practice of giving discounts, thus adversely affecting government revenues. Market prices, however, were in decline during the 1960s and so long as taxes were linked to actual revenues of the companies, government income would suffer. The government increasingly felt that the market price did not reflect the true value of petroleum. Dr. Pérez Alfonzo believed in an "intrinsic price" of petroleum, higher than the market price.

The government challenged the income tax returns of the companies on grounds of incorrect prices. Years of bargaining ensued in which the companies defended their returns and the government pressed for a different pricing formula. In October 1966, the successor to Dr. Pérez Alfonzo, Dr. Manuel Pérez Guerrero, announced the results of the negotiations. In effect, the government won its case. The companies recalculated back tax returns and paid additional money to the government. In the future, tax returns would employ reference prices rather than market prices. The negotiations had established the reference prices for 1967 through 1971. In 1970 congress gave the President the right to set reference prices unilaterally.[13]

Internationalization of Petroleum Policy

A favorite game of the oil companies was to play one country off against another. If one country demanded too much, the companies reduced their activities there and expanded in countries which were less demanding. This device, or the threat of it, was often effective in bringing the offending country into line, since its well-being (through exports, revenues and investment) depended on company actions. With international oil companies, individual national policies are ineffectual.

Mexico and Iran had felt the sting of the oil companies after their nationalizations. Venezuela felt it during the 1960s as its policies reduced company profitability. The companies disinvested, output stagnated, and reserves began to decline as the oil companies switched to the Middle East. To defend themselves against this sort of company behavior, oil exporting countries began to consider coordinated policies. During the 1950s several meetings among oil exporters, including Venezuela, touched on the possibility of an international producer organization.

The catalyst for action was the price decline of the late 1950s and early 1960s. The two principal architects were Dr. Juan Pablo Pérez Alfonzo of Venezuela, and Sheikh Abdullah Takiri of Saudi Arabia. In September 1960, five major oil exporting countries (Iran, Iraq, Kuwait, Saudi Arabia and Venezuela) met in Baghdad and formed the Organization of Petroleum Exporting Countries. Initially the OPEC countries attempted to persuade the oil companies to roll back prices. They were unsuccessful, but partly as a result of their efforts, price declines eased off.[14]

The oil companies did not take OPEC seriously. Experience had demonstrated the fragility of commodity agreements. The companies had had a great deal of practice in bilateral dealings with countries on oil issues and did not believe countries with such disparate interests could cooperate successfully in such a highly technical field. The early years of OPEC seemed to bear out the companies' judgment. With a glut in the petroleum market and companies' power at its zenith, OPEC did little but meet and talk.

Slowly at first, but with gathering momentum OPEC members learned the oil business. Membership grew to the point where OPEC members exported most of the world's crude oil. Members educated one another concerning petroleum policies, the behavior of the companies, and the petroleum business. Venezuelan policies were more advanced than those of the other countries, and for this reason during the 1960s Venezuela was a leader in pressing for uniform policies, greater revenue participation, conservation and high prices.

Delays, setbacks and controversies also beset OPEC. As the organization grew larger it became more heterogeneous. Some countries were conservative and felt that their interests were in placating the companies. Others, with less to lose or more to gain, wished to move faster. The oil companies skillfully took advantage of every difference. Most policies were still the policies of individual countries, backed up only with the moral support of the other members of OPEC.

By 1970 the rapid economic growth and lavish use of petroleum by the industrial countries had begun to turn the market, and pressure on prices began to mount. Some members of OPEC, notably Libya, made strong demands on the oil companies. In December 1970, at the meeting of OPEC in Caracas, the organization strengthened its position with respect to prices and participation, in part because of the Venezuelan congressional action on unilateral setting of reference prices. Talk of nationalizations, takeovers and embargoes emanated from oil exporters; in part reflecting continued uneasiness in Middle Eastern affairs, in part demonstrating growing confidence in the backing of OPEC in the emerging seller's market.

Each member of OPEC was influencing to a greater degree the activities of the oil companies in its own territory. Some demanded greater participation in revenue, some controlled production or investment. Increasingly, the prices the companies employed to pay royalties and taxes fell into the hands of the exporting country. The determination of market price, however, remained in the control of the companies, although the agreements between companies and countries were vesting more influence in the countries. Venezuela determined the reference price by executive decree.

In 1972 Venezuela tightened the screws a bit more with the reversion law. Using the 1943 petroleum law as the base, the new law specified in detail the conditions of reversion, which included the reversion of all properties of the companies. The companies also had to post bond to assure the government that the petroleum industry did not deteriorate.

Other members of OPEC demanded greater participation, by which they meant a greater share of ownership as well as more revenues. The increasing petroleum production of OPEC countries continued to lag behind the galloping demand of the West; prices began to rise. But the higher prices, occasional shortages, and continued uncertainty in the Middle East did not impel the West to reduce demand or take any energy measures.[15]

The 1973 Crisis and the OPEC Cartel

The petroleum crisis, the quadrupling of prices and the assertion of authority over prices by OPEC, originated in the outbreak of war in the Middle East area. While the Arab protagonists were not members of OPEC, the Arab nations which were members supported Egypt and Syria. Military resupply and money for Israel, as well as moral and policy support, by the United States and other nations offended the Arab nations and the Arab sub-group of OPEC announced an embargo. Venezuela did not join the embargo and the volume of its exports, which had been declining since 1970 went up briefly in the fourth quarter of 1973 to above the 1970 level, but declined again in 1974 and 1975.

In 1973 Venezuelan petroleum export prices stood 28 percent above the 1968–1971 level. As a result of OPEC's capture of petroleum prices in October 1973, when the oil companies (urged by OPEC to increase prices) refused to negotiate, the first quarter 1974 Venezuelan export price increased 298 percent and in April 1975, by another 31 percent. Compared to 1968–1971, Venezuelan petroleum prices had gone up five times by 1975. Production, on the other hand, was in 1975 less than two-thirds of 1970 production, the peak year.

OPEC became an intergovernmental cartel in late 1973 when it took charge of petroleum prices. The mechanism in the Middle East for controlling price is setting the price the oil companies must pay for the oil owned by the countries, and through taxes. The mechanism for Venezuela is the reference price. Both are imperfect mechanisms. If at the prices established by OPEC members crude oil accumlates in inventories, pressure on price mounts and price declines may ensue, despite the OPEC price. Only by reducing and regulating production also may OPEC countries effectively control price.

OPEC members have the power to control production. Middle Eastern and other OPEC members have nationalized the industry or control decision-making. Venezuela nationalized its petroleum industry in 1976. OPEC, however, has not yet worked out a production control mechanism, although voluntarily, in support of OPEC, many petroleum exporters have reduced production or restrained the rate of increase. The key to production control, of course, is Saudi Arabia and Iran, the largest producers. Venezuela's production has declined in recent years and in any case is small relative to total OPEC production.

Cartels are traditionally unstable and usually break up after a period of time. The reason for the establishment of a cartel is higher profits or greater growth. The producers combine because operating as a group enables each firm to make higher profits than it would if there were no group. Thus, the cartel, at least initially, mightily serves the interests of each. After the establishment of the cartel the benefits are less obvious. Suppose that it is an effective cartel, charging a monopoly price with profits divided up among producers according to their production quotas, and total production adjusts to the market to maintain the price. This assumed cartel is a more effective arrangement than exists in OPEC.

Under the postulated conditions it will soon occur to one of the members of the cartel that if it cheats on the cartel, it can make even more money. By producing and selling slightly more than its quota and charging perhaps a price slightly less, one firm can have the benefits of the cartel as well as additional profits derived from its betrayal of the cartel. The extra-cartel production and sales must be secret in order to avoid retaliation. Another firm will observe the same possibility for enhancing its profits. And then a third. Finally the cartel price breaks and the cartel disintegrates. A weakening demand or increased production by firms outside the cartel, of course, tends to accelerate the deterioration.

The OPEC arrangement faces the same possibility. Should the petroleum importing countries unite in a common policy and reduce consumption and imports, or if new discoveries outside OPEC (such as the North Sea, Alaska, China) add to production, OPEC could lose control of price. Should the exemplary cooperation among OPEC members falter and voluntary production restraint be insufficient, the petroleum importing countries could regain authority over price. Should individual members cheat on the system, OPEC price control could erode.

OPEC, however, is not a traditional cartel and other motivations may impart greater stability than those experienced by other cartels. The members are less developed countries whose economic motivation was to acquire funds for developmental purposes. The new prices, however, have produced revenues far beyond the capacity of these countries to absorb in the short-run and much of it has been recycled into international capital markets. Imports have increased, but not by nearly as much as exports. In 1974 Venezuelan exports increased 126 percent; imports 49 percent. Reserves of most OPEC countries are still very high. While all countries would like to have more reserves, their affluent position does not provide a strong motive for cheating on the cartel.

Political motives also color the OPEC cartel. The unresolved Middle Eastern conflict almost guarantees continued Arab solidarity and cooperation in OPEC. All OPEC countries feel that they have suffered economically and politically at the hands of petroleum importing countries and regard OPEC as an instrument of just retribution. Even non-petroleum exporting, less developed countries have aligned themselves with OPEC countries, against their own economic self-interest, in order to support the OPEC effort to reverse the politico-economic power imbalance.

The Venezuelan commitment to OPEC is strong. It is proud of its role as co-founder and supporter. OPEC has produced the revenue and if it continues, will produce more revenue for Venezuelan development. With reserves dwindling fast, except for the high-cost Orinoco Tar Belt, Venezuela feels that its best chance for diversification and development is within the OPEC framework.

Nationalization

Despite pressure from leftist political groups, the government of Venezuela did not seriously consider nationalization of the petroleum industry until 1973. At that time it became apparent that the financial burden of paying for the industry would not be onerous, that no disruption of the industry would occur, and that the task was within Venezuelan technical capability. In 1974 a special commission studied reversion and nationalization for the President, and in March 1975, the President proposed a nationalization law. It passed congress, and the President signed the nationalization law in August 1975.

The law itself does not provide the details of how the petroleum industry will operate under government ownership. It simply reserves the resource and the industry to the state. The one concession to private interests is a provision which permits the government to associate itself with the private sector under some circumstances and only with the permission of congress. The government wished for this provision, a source of heated debate, in order to acquire modern technology and for assistance in the marketing and distribution of petroleum outside the country.[16]

The country took over the companies integrally, leaving them as is, except that government officials instead of company officials occupy all decision-making positions. All companies will operate as before, except that the government established a holding company, Petróleos de Venezuela (Petroven), to oversee the activities of all companies. Over a period of time, Petroven will change the relationships among the public companies and the old companies, in accordance with an overall national petroleum plan. The president of Petroven, General Rafael Alonso Ravard, former head of the Corporación Venezolana de Guayana and one of Venezuela's finest administrators, has announced that his organization will operate free from political interference and as efficiently as had the private industry.[17]

NOTES

1. Aníbal R. Martínez. *Cronología del Petróleo Venezolano*, 1970, p. 22; Aníbal R. Martínez. *Historia Petrolera Venezolana*, 1973, p. 161.
2. Luís Vallenilla. *Auge, Declinación y Porvenir del Petróleo Venezolano*, 1973, pp. 77–90.

3. Edwin Lieuwin. *Petroleum in Venezuela*, 1954, pp. 70–100; Romulo Betancourt. *Venezuela: Política y Petróleo*, 1956, pp. 25–97.

4. Franklin Tugwell. *The Politics of Oil in Venezuela*, 1975, p. 41.

5. Aníbal R. Martínez. *Historia Petrolera Venezolana*, 1973, p. 162.

6. César Balestrini. *Los Precios del Petróleo y la Participación Fiscal de Venezuela*, 1974, pp. 35–41.

7. G. C. Hufbauer and others. *U. S. Taxation of American Business Abroad*, 1975, pp. 37–66.

8. Francisco Mieres. *El Petróleo y la Problemática Estructural Venezolana*, 1969, pp. 87–156; Michael Tanzer. *Political Economy of International Oil and the Underdeveloped Countries*, Boston: Beacon Press, 1969, pp. 9–40; Anthony Sampson. *The Seven Sisters*, pp. 43–112.

9. Aníbal R. Martínez. *Cronología del Petróleo Venezolano*, 1970, pp. 103–105; Franklin Tugwell. *The Politics of Oil in Venezuela*, 1975, p. 46.

10. Franklin Tugwell. *The Politics of Oil in Venezuela*, 1975, pp. 47–49; Luís Vallenilla. *Auge, Declinación y Porvenir del Petróleo Venezolano*, 1970, pp. 103–105; Franklin Tugwell. *The Politics of Oil in Venezuela*, 1975, pp. 47–49; Luís Vallenilla. *Auge, Declinación y Porvenir del Petróleo Venezolano*, 1973, pp. 208–222.

11. Pedro Esteban Mejía Alarcón. *La Industria del Petróleo en Venezuela*, 1972, pp. 121–123.

12. Franklin Tugwell. *The Politics of Oil in Venezuela*, 1975, pp. 174–175; *Resumen*, 21 December 1975, pp. 12–26.

13. Luís Vallenia. *Auge, Declinación y Porvenir del Petróleo Venezolano*, 1973, pp. 506–508.

14. Aníbal R. Martínez. *Nuestro Petróleo*, pp. 99–142.

15. James Akin. "The Oil Crisis: This Time the Wolf is Here." *Foreign Affairs*, Vol. 52, No. 3, Spring 1973, pp. 462–490.

16. *Resumen*, 21 December 1975, pp. 13–26.

17. *Resumen*, 4 January 1976, pp. 16–39.

Chapter V

Economic Policies

At the end of the Gómez regime economic policy served primarily private interests. The government collected taxes and spent money, regulated enterprise, and erratically intruded in the economy. These policies served the dictator by supporting and maintaining his control over the political and economic system, as well as an oligarchy of agricultural, commercial and financial families and enterprises favored by the dictator. Since 1936 Venezuela has installed economic policies appropriate to its mixed capitalist system, patterned after the policies of developed countries, and oriented to the achievement of economic growth, diversification, stability and justice.

The policies employed by Venezuela to reach these goals are the amounts and origin of taxes, the amounts and direction of expenditures, regulation of the financial system, policies toward the private sectors, international economic policies, and economic coordination and planning. The government has created many new institutions to carry out its policies and has intervened directly in the economy as well.

The catch phrase "sow the petroleum," coined by Dr. Arturo Uslar Pietri in 1936, defined the basic development strategy. Without altering basic economic and property arrangements, petroleum revenues, spent primarily on economic and social infrastructure, have spurred economic growth. The economic growth has changed the relationships among the sectors and between the private and public sectors, thus diversifying the economy and contributing to its stability. The increments to output promoted economic justice through their more equitable distribution.

Conflicts among goals have arisen and the mix of policies has changed as different governments emphasized different goals. Military regimes, especially the one between 1948 and 1958, chose to stress immediate economic growth, even at the expense of diversification, stability and justice. Democratic governments, on the other hand, have tended to focus more on the egalitarian goal and greater emphasis on changing sectoral relationships, with some sacrifice of growth, but without an all-out effort to change the nature of the economic system.

Tax Policy

Taxes in 1936 were a bewildering array of levies on goods and services, some dating from colonial days. The most important was the import duty, which was one-third of revenue. The second most important was petroleum royalties, officially not a tax but rather a payment for the alienation of national patrimony. In fact, it was an export tax. Other taxes, such as lighthouse and buoy taxes, dockage and cabotage fees, consular duties and others impinged on international trade. All trade-related taxes were 70 percent of government revenues. The rest were licenses, fees, stamp taxes and other assessments, a small part of which state and municipal governments collected.[1]

The prices of goods and services included these taxes. Since Venzuela exported most of its petroleum, Venezuelans paid little of the royalties and petroleum taxes. The other taxes Venezuelans paid when they bought goods and services. Indirect taxes such as these are regressive; all people pay the same tax regardless of their incomes. People with low incomes pay a larger share of their income in taxes than those with higher incomes. The two principal objectives of the tax policy of successive governments was the replacement of indirect taxes by direct and progressive taxes, and shifting more of the tax burden to foreigners through taxes on the export sector. In the 1970s indirect taxes were only 15 percent of government revenue, and the petroleum sector paid two-thirds or more of all revenues.

The first major reform in tax policy came in 1943 with the introduction of the progressive income tax. Greater revenue was its purpose, but it also shifted more of the tax burden to foreigners and reduced the share of regressive indirect taxes in total revenue. Nominally affecting all Venezuelan natural and juridical persons, the progressive income tax was in fact a tax on the oil companies since taxable income minimums and exemptions excluded most others. In a few years income taxes exceeded import duties and were growing faster than were royalties.

The government has changed tax rates and methods for calculating taxable incomes on many occasions. In earlier years the government endeavored to extract more and more from the oil companies with little consideration of domestic enterprises and individuals. While the government has never let up in its determination to maximize its revenues from the oil companies, during the 1960s and 1970s domestic taxation loomed larger in government policy. The government reduced minimum taxable income, tightened up exemptions and deductions, introduced a pay-as-you-go system for oil companies and salaried people, and made a more serious effort to collect domestic taxes. In part, this reflected a need for more revenues, but in part it also was a recognition of the limits and the long-term declining prospects of petroleum taxation and the need for a broad domestic tax base.

While the government did not fully articulate its petroleum tax policy, its aim was to capture all petroleum revenue above costs, plus only those profits necessary to keep the industry functioning. Dr. Juan Pablo Pérez Alfonzo believed these profits should not exceed 15 percent of invested capital.[2] The various tax and tax-related measures, exploration and surface taxes, income taxes, royalties, negotiated reference prices from 1967 to 1971 and unilateral reference prices thereafter, and changing tax rates all aspired to effect this limitation on what the government believed were unearned foreign factor payments.

Both domestic sectors and the petroleum companies have resisted taxes and employed every strategy to avoid them. The petroleum companies during the 1960s, took very high depreciation to raise their costs and thus reduce taxes. The move seemed like a good idea for the companies at the time, but in the nationalization of 1976 the government is paying net book value. In 1966 the government proposed a tax reform which simultaneously affected both the domestic and petroleum sectors. They formed an alliance which successfully stalled the reform, although it was at this time that the petroleum companies negotiated a settlement which introduced reference prices. In 1975, with the petroleum revenue question out of the way, with impending nationalization, the government renewed its reform of the domestic sector with measures which include higher rates and stricter enforcement.

Tax administration and enforcement has lagged. Individuals and enterprises often do not report their incomes or report false incomes and the government does not have sufficient independent information and enforcement personnel to challenge the data or call in non-reported income. One obstacle is a common sentiment among many Venezuelans that domestic taxation is inappropriate with huge oil revenues, an attitude often fostered by government inattention to domestic taxation. The proposed 1976 tax reforms aim to correct this impression, improve collections and convince people that taxes are an obligation of all.

Expenditure and Debt Policy

General Gómez believed in a balanced budget, no debt, and modest Treasury reserves. He achieved this goal; the last of the domestic debt he paid off in the early 1930s, and in 1935 the treasury balance was more than one-half of expenditures. Succeeding regimes have followed the same policy but have not been so faithful to its execution. Each year a balanced budget passes congress, although in recent years government income includes new indebtedness.[3]

In the early 1960s the government introduced budgeting-by-program in an effort to improve the efficiency of government expenditures. The budget specifies tasks for each governmental unit or category of expenditure and locks expenditures to the degree of fulfilment of the tasks. Venezuela was among the first in Latin America to install the system. While it has not revolutionized public spending, it has improved somewhat the spending process.

The government has not employed expenditures systematically to stabilize income. Some stabilization in practice resulted from incorrect government income estimates in the budget. If income went down expenditures exceeded income and treasury balances fell; if income went up, income exceeded expenditures and treasury balances accumulated. During the 1958–1962 recession the government incurred debt to maintain expenditures as a temporary stabilization measure. Upon recovery, Dr. Pérez Alfonzo suggested a stabilization fund from petroleum revenues to help stabilize the economy, but the government did not implement the plan. The present Fondo de Inversiones de Venezuela is not a stabilization fund but rather reflects the inability of the government to spend sudden large revenues on long-term development projects.

From the elimination of government debt in the early 1930s until 1958 the government did not incur any official foreign debt although a small internal debt accumulated. General Pérez Jiménez, however, permitted governmental units to contract short-term unofficial debt. The interim government and succeeding Betancourt presidency used the treasury accumulations from the 1956–1957 concession payments and borrowed in international markets in order to pay the contract debt.

When the economy recovered, the government began to employ debt as an instrument of policy. Slowly during the 1960s, but at an accelerating pace the government financed income-producing development projects with both foreign and domestic debt. The government made an effort to incur debt only when the project paid for itself. Even then debt required special justification by the planning agency and executive before congress, which must approve all foreign loans. Governmental units also use foreign loans as external validation for development projects.

The government has employed debt with great caution; in 1975 the outstanding debt was only Bs9,000 million. Less than one-third of it was foreign debt, and the foreign share was declining. Including the petroleum and iron ore nationalization, a 10-year debt of less than Bs5,000 million, total indebtedness was only about 20 percent of foreign exchange reserves in late 1975.[4]

Direction of Expenditures

The government directs expenditures primarily to the social sectors. Somewhat less than half of expenditures go to education, health and sanitation, social security, housing and other social services. The next largest group of expenditures is in the economic sectors, which absorb a little less than one-third of total expenditures. One-third of this spending goes to agriculture with the rest distributed among mining, manufacturing and construction. The remaining one-fourth of total expenditures finances general services, defense, interest on the debt, and the acquisition of financial assets.[5]

The construction of social and economic infrastructure, provision of increasing social services, and support of the private sector are the goals of expenditure policy. Direct public sector participation in the economy has been a lesser goal except in resources and some basic industries where the intent is complete government control and operation. Through social security and subsidized housing and food, Venezuela has a limited welfare program. The government itself is an amplification of the welfare program, as the employer of last resort. Many whose productivity is very low are on government payrolls.

In an effort to replace the wasting petroleum asset, the government has invested about one-third of its expenditures. This amount increased sharply when the petroleum revenues increased in 1974, and the government has laid aside a large amount in the Fondo de Inversiones de Venezuela for investment in large development projects, such as steel, aluminum and hydroelectric projects. In addition, much of the spending classified as "current" goes to improve productivity, as in the case of education, health, lending in the private sector, and housing.

Policy Toward the Private Sector

Governments of Venezuela have not challenged the right of the private sector to exist, expand and make profits. A tendency toward gradual encroachment on the private sector by the public sector is evident, however. The government has staked out a claim on petroleum and mining and related basic industries, such as steel, petrochemicals and aluminum. The government also participates directly in productive activities where the private sector is reluctant. But most activities in industry, commerce and services are in the private sector. The private sector, however, becomes relatively smaller as the government expands more rapidly in enterprise. The recent nationalizations have greatly enlarged the public sector.

The government has also restricted the private sector through the imposition of taxes, norms and codes of conduct, regulations and restrictions, and decrees affecting business behavior. A strain of the old Spanish paternalism still remains. The government endeavors to achieve policy goals through the private sector. For example, there are minimum wages and maximum prices. In 1974 the government directed that all washrooms and elevators have an attendant, and in 1975 that all employers increase the number of people on their payrolls.

The government has also promoted the development of the private sector. In agriculture, industry and commerce the government lends substantial sums for private enterprise to expand their activities or to initiate new ones. The government also provides capital by buying stock in private enterprise and by establishing mixed enterprises in which the government shares in decision-making, management and profits. Parts of industry and agriculture receive government subsidies. Private enterprise also maintains high profits through less-than-vigorous tax collection and lax enforcement of many restrictions.

The private sector provides the goods and services needed by the government, often on generous contract terms. The government is the biggest customer of many prosperous enterprises. Much of the government's expenditures directly benefit the private sector, such as those for highways, electric power, worker housing, training programs. The government builds industrial parks and gives them to private enterprise. Private enterprise in many sectors can compete with foreign goods because of the high protective tariff, and the government exonerates import duties on a wide range of imported goods used as inputs by enterprise. Limitations on foreign investment and restrictions on foreign enterprise also assist Venezuelan private enterprise.

A slowly moving equilibrium between the private and public sectors inches the economy toward government dominance. The government restricts, regulates, directs and encroaches on the private sector, bit by bit reducing the scope of private enterprise. At the same time it promotes, provides funds and pursues envigorating policies which keep the private sector prosperous and dynamic. No government policies portend complete government control over the economy, despite the quantum leap of petroleum nationalization, but the mix of policies presages the increasing importance of the public sector.[6]

Government Participation in Productive Activity

On a mounting scale the government has been an active participant in the economy, through the production and sale of goods and services. In some cases it has traditionally had a monopoly, as in the case of firearms, salt and internal communications. In other cases government participation derives from the basic government policy since 1959 of control of resources and resource-related industries, such as petroleum, mining, steel, hydroelectric power and petrochemicals. In still other cases, the government and private enterprise compete or supplement one another, as in banking and tourism. The government has been responsible for a significant and rising proportion of total production.

In addition to wholly owned and operated government, autonomous institutes and enterprises, the government also participates in various forms of mixed enterprises. In some cases the government is simply a minority shareholder and government interest is mainly to provide capital and assistance. In other cases the formation of a mixed enterprise signifies government determination to control, in part, the activities of the enterprise. The private and public sectors each bring needed elements to the mixture: management, technology, capital, markets, resources.

Three reasons impel direct government participation. If the private sector is unwilling or unable to initiate or expand production, the government may enter the field. For example, the private sector lacked the capital required to build the steel mill in the Guayana and the hydroelectric facilities on the Caroní. The second reason supported these government operations. When the government considers that the public or national interest is at stake, as in the case of resource and resource-related activities, the public sector exercises the initiative. The government felt that the Venzuelan flag must show in international air transport and thus formed, with KLM, the mixed company, VIASA. For the same reason, within the country, one of the two domestic airlines is government owned.

The third reason relates to the performance of public and semi-public functions. If, in the government's view, the private sector does not provide sufficient banking, marketing, tourist, or other facilities which serve the public, the government will perform these services directly.

Some government enterprises make profits and others suffer losses. The government usually provides the patrimony and in some cases operating resources from the national budget. Profits become income and losses become expenditures. From the government's point of view profit is not a reliable guide to the status of a government enterprise since the price of the product or service may, as a matter of policy, be low to encourage its use or to redistribute income. All government enterprises, however, must adhere to the regular norms of budgeting and accounting and are under the supervision of one of the ministries.

The acceleration of direct government participation in the economy in the last two decades reflects the government's development orientation and not a dogmatic stance contrary to private enterprise. The government perceives that the private sector, appropriately controlled, makes a valuable contribution to development. When a development need arises and is not met by the private sector, the government, on pragmatic grounds, does it, with little thought to ideology.[7]

Controls and Regulations

Venezuela has a regulated and paternalistic economy. The system of controls and regulations is so complex that both the regulators and the regulated get lost in the labyrinth of norms, standards, restrictions and decrees. Many are obsolete; others are unenforceable. Still others the government has never enforced or even tried to enforce. When established, a new regulation may have performed or proposed to perform a useful purpose. Time often eroded the regulation; enforcement was difficult or abandoned. Changing technology, products, practices rendered the regulation an anomaly. Businesses learned what regulations to obey, what to ignore.

The purpose of the regulatory mechanism has been to compel behavior which the government considered desirable for the public good. Some regulations have wide scope, applying to all enterprises or large groups of enterprises, such as the employment decrees of 1974 and 1975. Others are narrow, applying to only a few firms or only a particular class of firms. The automobile sector, for example, must use specific proportions of its components of domestic manufacture.

Two of the most important regulations are price and wage controls.

The government establishes maximum prices for most food products, medicines, fuels and rents, as well as for many other items of primary necessity. The purpose is to benefit low income people through low prices. The government also establishes minimum wages that vary with the nature of the work and conditions of work. Minimum prices on some agricultural products guarantee minimum income for agricultural workers. The government also, on occasion, establishes the numbers and kinds of workers in business enterprises.

Despite the plethora of regulations, the restrictions on Venezuelan enterprise are less dampening than in most economies. The recent wave of regulations which have inundated advanced countries concerning environmental protection has not yet hit Venezuela. Many regulations are little more than systematized common sense, appropriate to the beginning stages of modern enterprise. Standard business practices quickly absorb the regulations which then become innocuous. For both the regulated and the regulators, learning to operate industry and commerce efficiently and in the public interest is a trial-and-error process, and those regulations not absorbed or scrupulously enforced by the government fade away.[8]

Agricultural Policy

In addition to sharing amply in expenditures, agriculture has been a special target of government policy in the last four decades. Before the advent of petroleum, Venezuela had an agricultural economy and produced most of its foodstuffs. Coffee and cacao provided the foreign exchange. Even though agricultural resources were poor—limited arable land, heavy erosion, tropical climate—agriculture provided employment and livelihood for most Venzuelans and was the mainstay of the economy. Petroleum, in its early years, sucked labor and capital out of agriculture, weakening agriculture and permitting food imports that discouraged agricultural production.

Agricultural policy since 1935 has endeavored to fortify the sector by increasing both land in use and productivity, as well as by diversifying crops. More overhead capital, more working capital, changes in the land tenure system, and the introduction of modern technology and enterprise were and are the basic instruments of policies designed to create a growing responsive agriculture and agricultural self-sufficiency. Before the mid-1940s the principal aim was to recapture the land from the relatives of Gómez and introduce more capital into the sector.

The first move to reform the land tenure system came in the 1945–1948 period, but the succeeding military regime abandoned the policy and neglected agriculture. The Agricultural Reform Act of 1960 initiated a major effort to redeem the land and build a modern agricultural sector. The government gave land to campesinos from the public domain and from private owners, which it had purchased. The government built settlements, provided capital and technical assistance, plowed new roads through to isolated communities, and tried to educate the campesino.

With large government expenditures, a fresh breeze blew through agriculture. High protective tariffs and quotas guaranteed much of the domestic market to Venezuelan agriculturalists. The government-owned Banco Agrícola y Pecuario came to life and supported (with easy credit) the middle-sized, market-oriented farmers. The Instituto Agraria Nacional, in charge of the land reform program, promoted fertilizer use, experimental farming and land reclamation. The government built warehouses and storage facilities and improved the marketing facilities. While policy implementation wavered in the late 1960s and early 1970s, the policy outlook did not, and the government has earmarked a part of the increased revenues of the mid-1970s for agricultural development.[9]

Industrial Policy

Industrial policy before 1959 was largely unconscious. The private industrial sector responded to government expenditures, particularly in construction, and bountiful profits provided the stimulous to industrial growth. Monumental public works, road building, hydroelectric power plants, the beginning of the steel mill and petrochemicals benefitted industry, as did protection from imports in parts of manufacturing, but these were uncoordinated governmental activities, not an industrial policy. Indeed, in other respects, government damaged industry as in the case of the trade agreement with the United States, which opened the domestic market to imports and precluded Venezuelan production.

The industrial policy begun in 1959 has been openly expansionist. The new government initiated an import-substitute program, supported by higher tariffs and quotas, and the exoneration of customs duties on imported raw materials and intermediate and capital goods for domestic industry. Venezuela modified the trade agreement with the United States unilaterally in 1960 and abrogated it in 1972, at the same time it introduced an *ad valorem* tariff system to replace the old specific duties. In the import-substitute program the government directed its efforts toward replacing final goods, the first and easiest step, and has moved cautiously into efforts to replace imported intermediate and capital goods.

The bulk of industry remains in private hands, with government influence felt through its expenditures and through the provision of credit. The Corporación Venezolana de Fomento, launched in 1946 but languid until the early 1960s, lends substantial sums to industry and purchases stock in private companies in order to provide for their financing. In the 1960s and 1970s it has been a critical element in investment funds for industrial purposes, especially in new industries. The government-sponsored Banco Industrial and the national commission for small and medium-sized industry also made important contributions.

The government reserves its direct participation to resource-related industry, such as steel, aluminum, petrochemicals, petroleum refining and hydroelectric power. Even in these cases, the government invites private sector participation through mixed companies. The government has not been hospitable to private foreign investment in industry, fearing further penetration and dependence. It applies Decision 24 of the Cartegena, which spells out foreign investment opportunities and conditions in a restrictive fashion. In 1974 the government established an agency to monitor and regulate all foreign investment.[10]

Labor Policy

Labor policy has centered about two propositions: direct intervention in the labor market to improve the pay and working conditions of labor and to increase employment, and encouragement of labor organizations to effect the same goals and to countervail the influence of enterprise. Before 1936 labor unions were by definition communist and anathema; workers had no protection and could not strike. The first important labor law came in 1936, and as modified and improved in 1945, 1947 and 1966, is the basic labor code.

The original legislation permitted union organization and required union recognition. In 1937 the government created the labor ministry and in 1945 the labor law brought agriculture under its jurisdiction. The governments of Generals López Contreras and Medina Angarita countenanced the growing unionization of the labor force, although their governments intervened in strikes and interfered with labor meetings. Unions, however, began to bargain collectively.

The politicized labor unions provided much of the support for the military-AD takeover in 1945 and had great influence in the subsequent three-year government. The new government fomented union formation and the grouping of unions into regional and industrial federations. For the first time a central labor organization resulted from the government's efforts to solidify its labor support. The government encouraged collective bargaining and made it more effective. The policy came to an abrupt end with the 1948 military takeover. The new military regime suppressed nearly all union activity, drove the unions underground and halted the progress of labor.

After the ousting of the dictator in 1958, labor policy reverted to that pursued from 1945 to 1948. The government recognized all unions which had previously functioned, and encouraged the formation of additional unions. It restored collective bargaining and the right to strike. In addition to revitalizing the union movement, the government introduced measures to improve conditions for non-union workers, including regulation of hours and overtime, conditions of work, and minimum wages.

Since 1958 the government accepted the responsibility for the maintenance of full employment. During the recession it sponsored a make-work public works program to try to reduce unemployment. A part of the purpose of public expenditures was an effort to keep employment at high levels and the government itself has enlarged its own payroll to absorb workers. A modest labor placement service sought to find new jobs for the unemployed. In the mid-1970s the government began to use more forceful measures, requiring that employers expand the number of workers employed and that certain tasks make use of labor even when not necessary.

Another policy line hopes to improve labor productivity. The growing educational effort indeed has the effect of improving productivity and retaining young people in the educational system and out of the labor force. Specific programs in agriculture and industry seek to train workers and encourage on-the-job training by employers.[11]

Trade Policy

In 1935 Venezuelan trade policy consisted of import taxes (mainly specific duties) and petroleum royalties and minor levies, which were in fact export taxes. In the last forty years Venezuela has broadened its trade policy to encompass protective tariffs, *ad valorem* duties, bilateral and multilateral trade agreements, participation in international organizations, exchange rate controls, and regulation of foreign investment. The government has blended all of these policies in an effort to promote development and diversification and reduce dependence.

Import policy in 1935 was primarily for revenue purposes. As revenue from petroleum increased, and especially after the income tax in 1943, the revenue aspects of import policy faded and protection emerged as the stronger motive. The trade agreement with the United States, first negotiated in 1939, contained the most-favored-nation clause which, along with U.S. dominance of Venezuelan imports and exports, kept many import rates low in the interest of promoting and regularizing petroleum exports to the U.S. Venezuela maintained a number of non-tariff barriers, such as consular fees, which at one time had served a revenue function but gradually became nuisance taxes.

The trade agreement was renewed in 1952, again for the purpose of promoting exports and again with the effect of blunting the little protection provided by the specific duties. The democratic government which took office in 1959, however, decided to promote industrialization through the substitution of imported goods with domestic goods. The principal instrument was import policy. While still officially bound by the trade agreement, Venezuela employed the escape clause, arguing that the U.S. had also violated the agreement, and imposed greatly increased duties and introduced quotas and other non-tariff barriers on a wide range of agricultural and industrial products.

In 1972, when it was time to renegotiate the trade agreement, Venezuela renounced the agreement and embarked on a frankly protectionist policy, except for Latin American trading partners. The government renovated the entire import duty system, substituting *ad valorem* for specific duties and improving the administration of import policy significantly. Duties on final and semi-manufactured products are high, and lower on intermediate and capital goods. The government exonerates duties on many imported products which are inputs for domestic manufacturers, and those for development projects. Import policy now has a protective, sumptuary and development orientation.

Export policy has centered on expanding petroleum exports, diversifying the country of destination, and promoting non-traditional exports. Since the late 1930s most revenues originated in petroleum; the policy therefore promotes government revenues and spending, as well as providing the basis for imports. Venezuela has felt acutely its dependence on the U.S. market and has tried to move into other markets. At the same time it vigorously protested limitations placed on Venezuelan petroleum by the U.S. voluntary and mandatory quota policies of the 1950s and 1960s. While the government has officially endorsed promotion of other exports, only recently have subsidies and other measures been employed to make concrete its purposes.

Venezuela lagged behind other Latin American countries in participating in economic integration plans. The private sector feared competition in domestic markets from the more industrialized members of the Latin American Free Trade Association and the government did not forsee significant advantages for the petroleum sector. In 1966, however, over private sector objections Venezuela joined, although by that time national interest had triumphed over integration and LAFTA was almost moribund. The Andean Group replaced LAFTA as the center of the integration movement for the Andean countries but Venezuela's trade agreement with the U.S. barred its entry. When the agreement was abrogated, Venezuela joined the Andean Group. Venezuela has also strengthened its ties in the Caribbean area and after the 1973 OPEC success has begun to take an active leadership role in Latin American regional matters.[12]

Economic Planning

Before the 1958 interim military regime established the Central Office of Coordination and Planning (CORDIPLAN), Venezuela was innocent of economic planning. General Marcos Pérez Jiménez had a small study group to advise him, but he relied primarily on uncoordinated ministerial actions and a small coterie of insiders to help him in public spending. AD's dedication to planning was evident in the caliber of appointments and high priority attached to CORDIPLAN, whose principal function was to coordinate and rationalize expenditures. Planning, as conducted by CORDIPLAN, has been Venezuela's principal policy innovation.

The emphasis in CORDIPLAN has been on the first word: coordination. The generation of projects, technical and economic studies, and execution and supervision of projects remains with the ministries or other governmental units. CORDIPLAN functions as the filter through which expenditure proposals must pass on the way to inclusion in the budget and executive and congressional approval. In CORDIPLAN the small staff of planners examines the required benefit-cost studies and compares one unit's proposed activities with all others, seeking gaps, overlaps and uncoordinated activities of concern to two or more units.

CORDIPLAN's work is primarily technical and advisory only with respect to policy, since the President and his ministers set policy guidelines. In small working sessions with the concerned parties, CORDIPLAN, which is attached to the presidency, fills in, eliminates and matches expenditure proposals to fit into a coordinated plan which becomes, after the appropriate approvals, the mandatory public sector plan. The five-year plan outlines the principal features of the direction of government spending, and the one-year operational plans are the specific budget allocations year-by-year.[13]

CORDIPLAN is as important for what it is not as for what it is. It has never been a superministry designed to supplant decision-making by other governmental units, nor has it assumed a major policy role. Most of the planning falls to planning units within each governmental unit, and CORDIPLAN plays the role of educator, coordinator and catalyst for the work of others. CORDIPLAN does not have executive or policy powers but rather relies on persuasion and technical competence. CORDIPLAN is, in effect, the tip of the iceberg of planning and coordination.

CORDIPLAN has no authority over the private sector. It makes estimates of the composition and growth of the private economy in developing the public sector plan. In the early years the private sector feared that planning would clamp on it rigid controls and attempt to limit and direct its activities. The performance of CORDIPLAN has allayed these fears and has gradually drawn business and industrial leaders into the planning process. Indeed, the private sector has used CORDIPLAN's work as benchmarks in its own planning.

Venezuela employs "rolling" planning. CORDIPLAN is constantly planning and revising the planning as the execution of plans occurs. The government has published four major plans: 1960–1964, 1963–1966, 1965–1968 and 1970–1974. These documents, which summarize CORDIPLAN's estimates of the future course of the economy and the intentions of the government, are guides and reference works. The real essence of CORDIPLAN's work is the coordination of public expenditures as expressed in annual budgets. CORDIPLAN's written plans have been wide of the mark; in the 1960s because of lagging investment and poor petroleum performance, in the 1970s because of petroleum price increases. Degree of plan fulfillment, however, is not the appropriate measure of the success of CORDIPLAN.[14]

The proper measure of success in planning and coordination is the degree of improvement in economic decision-making, not the accuracy of plans. On that basis, Venezuelan planning policies have been successful. In the anti-planning Hispanic culture, the government bureaucracy accepts planning—and in an atmosphere of fierce individualism—coordination is becoming a way of life. While performance still falls far short of promise, the policy has begun not only to enhance economic decisions and the efficiency of the public sector, but also to provide the means for continuous self-improvement.

Early Financial Policies

The financial and monetary system of Venezuela in 1935 was primitive. General Gómez was not a great believer in banks; only six domestic and five foreign banks existed, all concentrated in Caracas and Maracaibo.

Limited central bank functions fell to the Banco de Venezuela, the largest commercial bank. The Treasury ministry had a small auxiliary bank and the government loaned money on occasion. No monetary policy existed and financial institutions received little notice or encouragement from the government.

The depression was still on; Venezuela had abandoned the gold standard. After the initial depreciation of the bolívar in the early 1930s, the bolívar had increased in value relative to the dollar. The balanced-budget, no-debt policy had reduced the liquidity of the financial system. Bankers were conservative, maintained high excess reserves and loaned only at short-term for commercial purposes to clients known to them. The banks operated like a private club, serving few clients, did not seek new business and did not compete. Almost no non-bank financial institutions existed. Banking and finance served the oligarchy without restraints from the government and without government encouragement to serve the broader interests of economic development.

In 1935 the petroleum companies and the banks, with a government minister sitting in, reached an agreement on a Bs3.93 to the dollar selling rate, but a lower buying rate for petroleum dollars. The international system was in chaos but the bolívar was strong and appreciated. In 1940 the buying rate was Bs3.09 to the dollar and the selling rate was Bs3.19. At these exchange rates non-petroleum exporters could not make ends meet, so that during the war years these exporters enjoyed a higher rate, a preference largely eliminated in the 1960s.

In the Gómez era private banks were custodians of gold and foreign exchange. In 1937 the government established an authority to capture and dispense exchange as needed by the banks. This was the beginning of government central banking. Through the late 1930s debate over a central bank raged. Private banks opposed the idea, which would eliminate their profitable government accounts and bill emission business. The government pressed the view that without a central bank the government had no voice in monetary affairs. It prevailed; in 1940 the fledgling bank began to function, a poor thing with little authority or flexibility. It did, however, centralize coin and bill emission, serve as the government's bank, and handle foreign exchange.[15]

Monetary Policy

Monetary policy is the manipulation of money supply and liquidity to influence the rate of interest and conditions of credit, thereby encouraging or restraining the economy. Monetary policy is also the degree of sophistication of the financial system and its efficacy in transferring monetary resources from savers to investors. On both counts, Venezuela begins with almost no monetary policy, but during the past forty years the country has gradually developed increasing control of the money supply and fomented an institutional arrangement, subject to government control, which assists saver and investor alike.

Money supply comes from private, governmental and international credit. The government levies taxes and spends the proceeds. When an oil company, a domestic enterprise, or an individual pays taxes, money in circulation or in the banks moves into government coffers. If the government spends exactly the same amount, the money moves back into circulation or into banks, and the transaction is a simple transfer; no change in money supply occurs. If the government spends more than it receives, it goes into debt by creating more money or deposits than it can liquidate; the money supply goes up. If the government spends less than tax payers turn over, then the amount of money in the hands of the public goes down and the government holds a reserve. By spending less than it taxes, the government reduces the liquidity of the monetary system and tightens credit conditions. By overspending, the government increases the money supply. A neutral fiscal policy—a balanced budget—implies that money supply depends on taxes which in turn depends on international transactions.

Venezuelan exporters sell their products and earn foreign exchange. Since 1937 the foreign exchange goes to a central institution and that institution—the Central Bank since 1940—creates bolívar deposits at a fixed exchange rate in return for the dollars. Foreigners investing in Venezuela likewise convert foreign money into bolívar deposits. All these transactions which involve the acquisition of foreign exchange in return for bolívares add to the money supply. On the other hand, importers and Venezuelan investors abroad surrender bolívares to obtain foreign exchange, thus reducing the money supply. Historically, the government has had virtually no control over international transactions and hence over this source of change in the domestic money supply.

Taxes link the fiscal and international sources of liquidity together. Before the 1940s taxes depended on import duties and taxes on petroleum exports. Subsequently, taxes originated primarily in the export sector. Starting with balanced government and international accounts, a decline in exports initiates changes in money supply, the direction and magnitude of which depends on government spending and imports. Domestic money supply thus depends, on both accounts, on international transactions.

The third source of money is private credit. If an enterprise borrows to finance inventory, the bank creates a deposit, lending out of excess reserves. The newly created deposit, since only a part of it need be held as reserves, provides the basis for another loan. A fractional reserve banking system which seeks to eliminate excess reserves will expand or contract in response to changes in deposits from outside the system. If new deposits come in on the government and international accounts, the banking system will tend to amplify them. If these accounts go down, the banking system will contract to accommodate itself to the change.

Even under the best of conditions it is extraordinarily difficult to have an effective monetary policy when the international account calls the tune for money supply and the banking system maintains large excess reserves. Monetary policy can influence reserve creation and destruction from within the banking system itself but cannot determine taxes and spending or imports and exports, which also add and subtract reserves. Monetary policy also loses force if banks do not respond to changes in reserves. Venezuela has fought an uphill battle to install an efficient monetary policy with only partial success.[16]

The Central Bank

The key to monetary policy is the Central Bank. It can set reserve requirements, specify its own lending conditions, buy and sell financial assets, directly control monetary conditions, and influence banks through its public position on the state of the economy. However, even in developed countries with finely-honed financial systems, great controversy surrounds the effectiveness of money policy. In Venezuela, with a less sophisticated system, external dependence and institutional barriers, few argue that monetary policy is a major element in the regulation of the economy. Even so, the Central Bank has done a remarkable job of controlling monetary conditions to some degree and in creating a financial system which is subject to monetary controls.

The Central Bank of Venezuela had few monetary policy functions at its birth in 1940. It centralized foreign exchange and currency emission, standardized banking practices, monitored reserves which were far in excess of those required and presented an image of monetary responsibility. Over the years, and especially since 1960, it has pressed for and acquired more authority, led in introducing new and varied financial institutions, become the principal source of economic data and analysis, and earned a reputation for soundness and accountability.

Monetary policy, however, failed its first real test in the recession of 1958–1962. The Central Bank, still with limited powers, admirably came to the rescue of the banking system and no banks failed, although private credit collapsed. The fiscal and international effects swamped the private credit and had an adverse effect on it. Only vigorous Central Bank advances and rediscounting prevented a catastrophe. The experience provided the basis of monetary reform which included increased power to the Central Bank and more diversified financial institutions and assets.

Under the watchful eye of the Central Bank the financial system of Venezuela has developed into one of the strongest in Latin America. In 1971 the government suppressed foreign banking. Government banking operations came under the scrutiny of the Central Bank. Reforms in the stock market provided benefits to listed companies and the investors. Savings and loan associations and home mortgage banks receive encouragement and benefits as well. In the 1975 banking reforms the Central Bank acquired the full panoply of monetary policy authority, including control over asset composition of commercial banks. While monetary policy still suffers from debilitating dependence on external factors and fiscal policy, the financial system, now large and diversified, has been domesticated.[17]

International Finance

Venezuela was a founding member of the International Monetary Fund, the International Bank for Reconstruction and Development, and the Interamerican Development Bank. It accepted the rules of the game of the international monetary system except for a special exchange rate to favor agriculture and benefit from petroleum. The government deployed selective exchange controls, administered by the Central Bank during the recession, and in 1964 devalued the special petroleum rate. In 1971 and 1973 the bolívar remained fixed to the dollar but on each occasion the government revalued slightly. The bolívar is a hard currency and Venezuela's reputation in international finance is impeccable.

When the great flood of financial resources began in late 1973, Venezuela restrained the domestic financial system by sequestering funds in a special account—the Venezuelan Investment Fund—and invested the funds in international capital markets. Venezuela has been a leader in recycling oil money back into the system, through enlarged contributions to international organizations, bilateral aid to Latin American countries, and investments abroad.

Ideology, Politics and Policy

The military regimes from 1935 to 1945, and 1948 to 1958 had no ideology or politics. Their basic principle was to stay in power and to benefit from that power. The 1952–1958 dictatorship wrapped itself in the flag and proclaimed the New National Ideal, but it was agnostic as to right or left and immune to consideration of goals other than those of the dictator. It was close to pure self interest. The regime accepted the private sector contention that rapid growth was a function of policies favoring business. At the same time it initiated giant government projects.

Acción Democrática inherited a left-of-center economic and social philosophy from its political past, and assiduously promoted it. Its outlook accepted the sanctity of private property and legal change, even though filled with revolutionary rhetoric. Its goals were development through growth, structural change, and the elimination of economic and social disparities; its policies promoted these goals gradually and within the existing social and economic framework.

Nonetheless, great controversy has surrounded nearly every policy move of AD, as well as COPEI. Contending parties politicized nearly every aspect of economic policy. The opposition attacked a policy of the government not because it opposed the policy but because it was the government's policy. The open press stimulated and heightened the debate and sought out differences to foment more squabbling. Leftists maintained constant pressure for more drastic policies, and the business community argued that proposed policies would lead to the collapse of the economy and demise of incentives. The next election influenced policy formation. Delay was the fate of policies which might be unpopular at the polls; hasty and ill-considered policies favored by the electorate found support everywhere, each party claiming them as its policies.

Despite all the political bickering and backbiting, the underlying framework of policy and the ideological outlook, not only of AD but also of other major parties, has remained constant. The changes in governments did not represent major changes in strategy, in policy, or in the machinery of policy-making. Each party has a different emphasis in policy but no important political leader suggested a reversion to the maximum growth policies of the 1950s or a radically modified role of the government in the economy.

Many charge that political parties spend most of their time and energies in seeking and winning power to the neglect of detailed attention to economic policy. Venezuela has not produced a master economic plan, and nearly all policies are temporary expedients to deal with specific evolving situations. The political system which generates policy is inefficient and produces congeries of ill-timed and unwise policies, along with a few good ones. In these respects Venezuela now resembles the great democracies of North America and Western Europe, with the added disability of greater external dependence.

Any politico-economic system which chooses representative government and mixed capitalism, as Venezuela has, pays a high price, resulting from economic mismanagement. The vicissitudes of politics will undermine sound policies, promote imprudent policies, and interfere with policy formation. Still, over recent decades Venezuelan goals have been remarkably consistent and policies appropriate to the goals. The grand design of economic policy and the specific intent of particular policies have been economic independence, diversification, socio-economic improvement and growth, and the economy has responded in kind.

NOTES

1. *Anuario Estadístico 1938*, pp. 422–459.
2. Franklin Tugwell. *The Politics of Oil in Venezuela*, 1975, p. 53.
3. The economic policy of General Gómez was rudimentary. He ran Venezuela as though he owned it. Two recent books throw light on the Gómez period: Domingo Alberto Rangel. *Gómez, el Amo del Poder*, Valencia: Vadell Hermanos, 1975; Luís Cordero Velásquez. *Gómez y las Fuerzas Vivas*, Caracas: Editorial DONeMe, 1971.
4. *International Financial Statistics*, Vol. XXVIV, No. 1, January 1976, pp. 406–409.
5. *Anuario Estadístico 1971*, pp. 767–790.
6. Domingo Alberto Rangel. *La Oligarquía del Dinero*, 1972.
7. Statement of the President, *Resumen*, 11 January 1976, pp. 6–8.
8. Luís Vallenilla. *Oil: The Making of a New Economic Order*, New York: McGraw Hill, 1975, pp. 192–196.
9. Chi-Yi Chen. *Economía Social de Trabajo*, Caracas: Librería Editorial Salesiana, 1969, Chapter XI.
10. Luís Vallenilla. *Oil: The Making of a New Economic Order*, 1975, Chapters 10 and 11.
11. Fred D. Levy, Jr. *Economic Planning in Venezuela*, New York: Praeger, 1968; *La Planificación en Venezuela*, Oficina Central de Coordinación y Planificacion, Caracas, 1973.
12. José González-Gorrondona, h. *La Planificación Económica y su Aplicabilidad al Caso Venezolano*, Caracas: Ediciones del Cuatricentenario de Caracas, 1967.
13. Alfonso Espinosa. *El Proceso Monetario: Venezuela 1930–1960*, Caracas: Editorial Arte, 1963.
14. Alfredo Machado Gómez. *Crisis y Recuperación*, Caracas: Banco Central de Venezuela, 1972, chapter III.

15. *Informe Económico 1974*, Banco Central de Venezuela, chapter II; *Revista del Banco Central de Venezuela*, October-December 1974, pp. 5–46.

16. *IMF Survey*, International Monetary Fund, 9 June 1975, pp. 174–176.

17. Daniel H. Levine. *Conflict and Political Change in Venezuela*, Princeton: Princeton University Press, 1975, chapters IX and X.

Chapter VI

Economic Growth and Structural Change

The Venezuelan economy has grown at a vertiginous rate in the last four decades. In 1935 Venezuela was one of the poorest countries in Latin America; in 1975 output per person was higher than any country in Latin America and rivalled that of some developed countries. In between, the economy experienced average real growth of seven percent per year, double that of the population growth rate. Compared to any other country for a similar period, the performance of the Venezuelan economy has been awesome.

Economic growth has been irregular, however. In good years growth exceeded 10 per cent, as in 1951, 1956–1957, 1964 and 1972, and after. In other years (1958, 1961 and from 1965 to 1969) growth was only slightly in excess of population growth. The pace of the economy has followed closely Venezuela's petroleum market. In the 1950s and 1970s when petroleum demand rode high, the economy zipped along rapidly. The petroleum price break in 1958 and the relatively stagnant petroleum sector in the 1960s resulted in a recession and sluggish growth through the decade.

Structural change accompanied growth. After years of relative decline, agriculture stabilized as a portion of the economy. Petroleum was the leading sector until the late 1950s, lost the position through slow growth in the 1960s, but regained it through price increases in the petroleum boom of the 1970s. Industry gained, relative to the economy, and commerce and services demonstrated impressive advances. The labor share of national income increased but the distribution of income among social and economic classes remained highly unequal.

Economic Growth, 1936–1950

Despite depression and war, the Venezuelan economy grew rapidly in the fifeteen years following the death of General Gómez. An important petroleum industry was in place in 1935; Venezuela was the largest exporter and second largest producer. The depression had reduced exports, but by 1933 the trend was upward again and petroleum in 1935 was 91 percent of total exports. Government revenues were also going up slightly but were still below the 1929 level. While the economy was growing again, no bouyancy of prosperity prevailed; everyone was still in shock that the depression could happen. Venezuelans awaited with uncertainty the death of the dictator.

No national accounts exist before 1950 except for a sketchy study of national income for 1936. That study reveals a national income of Bsl,499 million. With a population of 3.4 million, per capita income was Bs441. At the official exchange rate, that is $143. Prices were low; a pound of beef cost a little more than 15 cents, and sugar and rice were less than a dime a pound.[1]

National income in 1950 was Bs8,607 million or Bs1,702 per capita. Prices, however, had risen by about 73 percent, so that with 1936 prices per capita, national income in 1950 was 2.2 times greater than in 1936. Other measures also indicate growth. Government income in 1950 was 8.7 times greater than in 1936 in money terms and 5.0 times greater in real terms. Crude production was up 3.5 times and the real value of petroleum exports climbed 2.8 times.[2]

The growth path, however, was not smooth, largely because of the war. Crude production dipped in 1940 and declined sharply in 1942. In the post-war period to 1950 the increase was 69 percent. Increases in the production of refined products went up 10 times between 1938 and 1950. Many items not produced in 1936 began and expanded rapidly; paint production started in 1945 and increased 3.5 times by 1950. Thus was stimulated many lines of domestic manufacture. Despite the lumpiness of growth, by 1950 the economy began to resemble a modern economy.

Venezuela laid some of the foundations for later growth in this period. Agriculture received more attention and by 1950 was not in the earlier depressed condition; some of the rural labor force migrated to the cities. Petroleum was clearly in command of the economy and was the leading sector. The labor force had more skills. Capital formation surged ahead, not only in petroleum and petroleum refining, but also in domestic sectors, helped along by the modest wartime industrialization and the growth of an entrepreneurial group.

The public sector was especially vigorous in the 1936–1950 period. Government revenues in 1936 were Bs221 million and rose to Bs1,917 million in 1950, which was down slightly from the year before. The largest increases resulted from the introduction of the income tax and from the 1943 Hydrocarbon law. The increase in public revenue from 1943 to 1950 was 564 percent. The government spent all the revenues. Expenditures, therefore, for the domestic economy originated externally with the petroleum buyers. Venezuela increasingly made its resource pay for development, turning an external factor to the country's economic advantage.[3]

The matrix of economic, political and social conditions had changed for the better. In 1950 the beginning of the institutional framework for economic development was in place, including a Central Bank and more adequately organized government and enterprise. While the country was back in the hands of a military junta in 1950, it was nothing like the dictatorship before 1936; popular sentiments were playing a major role and many of the reforms of the 1945–1948 democratic experiment were still in effect.

By 1950 the people, the government, private enterprise, and the intellectual community accepted the development mystique. A decade and a half of growth had convinced Venezuelans that the country could develop. In 1935 Venezuela was at the bottom of the ladder, hardly daring to look up; in 1950 Venezuela was well up on the ladder, had gained confidence, and was taking surer strides upward.

Economic Position in 1950

The four outstanding characteristics of the Venezuelan economy in 1950 were: dependence on the petroleum sector, sensitivity to international markets, socio-economic distortions, and the absence of regional integration. The economic structure of the country was lopsided. A large part of total product was in the petroleum sector which accounted for nearly all exports. Expenditures of the government had expanded with petroleum revenues, and commerce, industry and construction danced to the tune of the petroleum companies and the government.[4]

High costs, which encouraged imports, impeded agricultural production. Some arable land was not in use and other land useful for crops was in natural pasture. Aside from modern methods on a few commercial agricultural estates, techniques were backward, rendering the country non-competitive in export markets and vulnerable to import competition. Manufacturing was small scale, limited to a few industries, and heavily protected. Managers and skilled workers were high cost and inexperienced, and industry was inefficient.

The economy was highly sensitive to international markets, particularly the petroleum market. Petroleum costs were higher than in the Middle East which was capturing Venezuela's markets. Exports permitted high imports on which the country had come to depend. Domestic agriculture and industry could not compete with imports, especially in view of the trade agreement with the United States, which left some of the domestic market open. The concentration of income in coastal areas, where access to foreign markets was easier than access to the interior, supported the preference for imported goods. Income inequality also sustained imports, since the relatively wealthy of the cities spent much of their income on imports, either because no domestic production existed or because of a preference for foreign goods.

Distortions persisted in the geographic pattern of population and production. The petroleum sector had attracted workers from the agricultural interior to the oil fields and surrounding communities. The government had spent most of its revenue in the cities, especially Caracas. Communications with the interior were poor. The economy was in two parts. One half the population, living in cities near the coast, had high incomes, better transportation and communications, and more adequate commerce and social services. The other half still lived in the countryside (many in subsistence agriculture), had lower incomes and few of the city amenities. People were migrating from rural to urban areas but integration between the two zones was poor, nor was there integration within rural areas.

The social structure reflected both the stratified society of colonial and nineteenth century Venezuela and the economic and regional asymmetry of the twentieth century. Less than one percent of land holdings accounted for two-thirds of the productive area in 1950. Per capita income was $550 but its distribution was highly uneven, with most people having incomes of less than $100. Housing was primitive. Unemployment and disguised unemployment were common. More than one-half of adults could not read or write and the majority of the children did not go beyond the first grade. In cities incomes were higher but extremes of poverty and great wealth prevailed. Those recently arrived from the countryside lived in cardboard shacks with no water or electricity while the favored few lived in sumptuous homes and luxury apartments.

In 1950 Venezuela was an underdeveloped country, divided into non-integrated economic, geographic and social sectors, and critically dependent on decision-making beyond the country's control. Compared to other countries it appeared to be better off because of higher average incomes but this was deceptive because of its inequality. Compared to its situation in 1935, however, Venezuela was vastly better off; economic development had begun, and the people and the government had accepted the challenge of development.

Economic Growth, 1950–1957

Venezuela registered one of its highest growth rates—nearly 12 percent for gross territorial product in constant prices—in 1950–1951. The Korean war gulped down huge amounts of oil, and world tensions were high; crude production increased 14 percent. The threatening stance of the 1945–1948 government toward the oil companies and its reform policies were in the past; a military junta, friendly to oil interests, ruled.[5]

The following two years, however, witnessed a decline in the rate of growth to about 6 percent. The Korean war wound down; an armistice was signed in mid-1953. Crude exports, which had increased greatly in 1950 and 1951, stabilized and even declined in 1953, although crude prices, long stable, began to advance. New refineries also came on line. The government crushed the labor unions and in 1952 General Pérez Jiménez declared himself president, dissolving the junta. The outlook for petroleum and for the economy remained favorable.

In 1954 and 1955 the rate of growth climbed once again, initiating an acceleration which lasted through 1957. The oil boom was on, as volume and value of exports mounted each year. Public revenues also grew rapidly, and public and private construction expanded frantically. The rate of growth of gross territorial product was more than 10 percent in 1956, nearly 12 percent in 1957; petroleum exports grew even more rapidly. The Middle East exploded in 1956; a month later the government granted new concessions. The dictator's spending spree exceeded revenues and short-term credit; in anticipation of expanding revenues he financed many projects. Every part of the economy felt the impact of oil money and debt as it circulated through the system.

Middle Eastern uncertainties continued in 1957 and the government granted more concessions. The price of crude advanced nearly 13 percent. There seemed to be no end to the largess of petroleum. But the glimmerings of trouble began to appear. The United States started to restrict imports, fearing a domestic surplus. New discoveries in the Middle East and Africa changed the supply outlook and Venezuela's market situation deteriorated, as its crude was higher cost and supertankers cut the transport cost advantage.

The year 1957 marked the end of an era for petroleum and for Venezuela. Petroleum exports grew 13 percent in 1957, fell by 6 percent in 1958.[6] In January 1958, the people took to the streets and demonstrated their outrage against the abuses of the dictator. Business, labor and military leaders conspired to depose the dictator. Despite ten years of prosperity, the government had not addressed, much less resolved a single deficiency in the economy. Economic, geographic and social distortion and disintegration, as well as dependence, had heightened. The economic progress of oil and dictatorship was a phantom and economic underdevelopment still reigned.

In 1957 the economy was operating at a higher level. National income that year was 94.9 percent greater than in 1950, and almost no inflation had occurred. Per capita national income had increased 56.7 percent in the seven years. But when, in 1958, Venezuelans began to recapture their country, external factors and the conditions of the economy were unfavorable. The economy still depended on petroleum production, which made possible imports that were essential in food and manufactured goods consumption. The larger industry was still anemic and inefficient and had not broken out of light consumer goods production.

Migration to the cities had continued, and housing and sanitary conditions deteriorated. Labor shortages existed in rural areas and land in use did not expand. The standard of living of the rural worker declined. Professor George Hill made surveys in 1945 and again in 1957. In 1958 he reported that "agricultural workers were still living in the same stage of ignorance and poverty as when we first met them. From the economic and social standpoints, we think it may be said that their situation is worse today than formerly."[7] The same was true of the growing army of urban poor.

Nearly one-half of the nation's private income went to people in Caracas which had one-sixth of the population. The inequality of incomes was greater in 1957 than in 1950. Urban wage rates had risen slightly, attracting more people to cities where slums, sickness, child abandonment and crime was mounting. Growth had benefitted primarily the urban middle and upper classes without correcting the deficiencies of the economy.[8]

Between 1950 and 1957 the government collected Bs24,522 in taxes, more than all the revenues from 1498 to 1950. The government spent even more by contracting short-term debt. The government wasted money by allowing high supplier and contractor profits, and through corruption and administrative irregularities. It made no effort to economize and undertook no project evaluations. Expenditures went into superhighways, hotels, office buildings, apartment blocks and monuments, mainly in Caracas. The economy in 1957 was bigger, had more of the façade of modernity, but was not better than in 1950.[9]

Recession Years, 1958–1962

The year 1958 was a troublesome and uncertain one. The interim government, a coalition of military and business leaders supported by all parties was in power, while an intense electoral campaign unfolded. The private sector hesitated. Crude production and exports dipped and the price of crude declined by more than 4 percent. Mining production stood still, as did agriculture. Industrial production made little progress. The economy which had grown at nearly 12 percent in 1957 grew only a little more than one percent in 1958. Unemployment, which had been creeping up since 1954, reached 9 percent.

The interim regime took only emergency measures, such as giving money away in the streets. It began to pay the contract debt General Pérez Jiménez had piled up, using Treasury reserves still swollen by the 1956–1957 concession payments. Foreign capital became jittery, international reserves went down, and the balance of payments capital account turned negative. Domestic investment declined and the economy seemed to grind to a halt. The government increased petroleum taxes, hoping to relieve the anticipated deficits.

In January 1959, Rómulo Betancourt and Acción Democrática returned to power. AD had promised a reform government and needed money. The government confirmed the new petroleum taxes and sought new ways of extracting money from the petroleum companies, including reference pricing. At first the government tried to hold the line on expenditures but the pressure was too great and the government felt the contract debt had to be paid. Capital flowed out and the balance of payments deficit continued despite declining imports because of new restrictions.[10]

The economy resumed growth in 1959. The price of petroleum, however, continued to go down, plummeting in 1959 nearly 12 percent. For the second year in a row petroleum export value declined, although the volume of crude increased slightly. The government's revenue increased in 1959, largely due to the increased taxes, and expenditures remained enormous, resulting in a substantial deficit. Domestic industry and agriculture began to grow as foreign goods disappeared, and commerce and services, also buoyed up by the government deficit, moved ahead strongly. Gross territorial product increased about 8 percent but unemployment continued to go up. The new government had prevented economic disaster but the economy was still weak and in recession.

The petroleum sector and much of the business community felt threatened by some of the AD reforms. Investment wavered. The banking system malfunctioned. In 1959 current account deposits declined and in 1960 both current and savings deposits dropped. Private lending collapsed. Capital began to flow out of the country on an even larger scale. In 1960 investment declined 22 percent; hardest hit were petroleum, manufacturing and transportation. The recession was in full swing in 1960. That year gross territorial product in constant prices rose by only 4 percent, but national income in current prices went down by 2 percent. Unemployment climbed to nearly 13 percent with Caracas and other cities feeling the heaviest burden. The price of petroleum went down by another 5 percent in 1960. Terrorism threatened the political system and great uncertainty clouded the economic and political future of the country.

The year 1961 was another recession year. Manufacturing made some progress but agriculture, which had grown in 1960, stagnated in 1961. Unemployment rose, investment declined. Government deficits continued. In the face of the dwindling reserves, the government introduced exchange controls. The recovery began in 1962, when investment increased and the banking system began to recuperate. Gross territorial product rose 9 percent, but the recession lingered on in unemployment which rose again in 1962. In the four worst recession years, 1958–1961, gross territorial product rose less than 5 percent per year.

Economic Growth, 1962–1970

The irregular growth from 1962–1965 averaged 8 percent per year. In 1962 it was high, declined in 1963, and then in 1964 shot up to almost 10 percent, the highest recorded in the decade. Then in 1965 the rate of growth slumped again. Unemployment continued to be a serious problem, although the economy had recovered. The highest unemployment rate was in 1962, the highest unemployment in 1963. Unemployment ran 12 percent for the four years combined, and in 1965 it was still nearly 9 percent. The price of petroleum declined slightly although production edged up. Export earnings from petroleum remained almost constant. The capital flight halted, import restrictions restrained imports, and international reserves began to build again.

Investment recovered and increased each year, with a large increase in 1964. Gross investment in petroleum, however, was modest, running one-fourth of the amounts in the heydays of the 1950s. Foreign investment did not even cover depreciation, and accumulated net foreign investment was declining. Investment in manufacturing declined in 1963 but accelerated in 1964 and 1965. Housing investment increased nearly three-fourths between 1962 and 1965, as government programs got under way. The ratio of investment to gross territorial product was substantially below the ratio of the 1950s.

While the rates of growth did not match those of the 1950–1957 period, the economy at least recovered fully from the recession. The difference between the mid-1950s and mid-1960s, and the principal weakness of the latter period, was the failure of petroleum to recover and lead the economy. Petroleum was following the economy; industry, agriculture, commerce and services were all doing better. In part, this was due to flagging demand for petroleum and the great outpouring of Middle Eastern oils. In part, it is also attributable to United States import restrictions introduced in 1958.

The high growth of 1964 was followed by a decline to a 6 percent rate of growth in 1965. Had the past been a guide, in 1966 the rate of growth should have increased; only in 1952 and 1953 had there been two succeeding years of decline in the growth rate. But in 1966 the rate of growth plunged again to the dangerously low 2 percent, only the second time real economic growth was below population growth. This was the year of the confrontation between the government and private and petroleum sectors.

While the government and petroleum and business debated tax and trade policy, the economy hesitated. Petroleum exports went down and price declined slightly. Agriculture grew very little. Manufacturing stood still. Even tertiary production, which had led the economy, made small progress. Investment stagnated. Even the banking system stuttered as deposits declined and international reserves went down. Despite the increase in public expenditures, 1966 was a bad year.

The year 1967 was better, but not by much. Gross territorial product grew 4 percent; unemployment remained at slightly less than 9 percent. All sectors edged up slightly. Petroleum exports increased nearly 6 percent, but another price decline eroded exchange earnings. With 1959 was the base, the terms of trade in 1967 were 83, with most of the fall attributable to lower export prices. All of the economic indicators in 1967 were up but the economy did not have the bouyancy of the earlier days when petroleum was surging ahead.

In 1968 the economy continued to make progress; the rate of growth was 5 percent, about average for the 1958–1970 period but far below the 1950–1957 period. The leading sector was services, and the relative position of petroleum slipped. Gross territorial product increased in 1969 by only 4 percent, and in 1970 by 6 percent. The economy was moving forward in these years slowly, under its own steam, with little help from petroleum.

The only way to characterize 1966 to 1970 is stagnation. The average rate of growth was about 4 percent, only slightly ahead of population growth. Government expenditures, however, were rising by about 6 percent per year. Petroleum exports stagnated and the contribution of petroleum to total product declined. At the end of the decade invesment in petroleum recovered, primarily because of the desulphurication plants, but investment in manufacturing was constant. Government investment increased, especially in housing and transportation.

The performance of the Venezuelan economy from 1950 to 1970 demonstrates a downward trend in the rate of growth, with 1958 as the dividing line between high and low growth rates. For 1950–1957, the average rate was more than 9 percent, and for 1958–1970 it was somewhat more than 5 percent. A trend line through the two periods shows a growth rate that declines one percent every three to four years. A prediction made in 1970 without taking into account drastic changes in the external sector, would show a continuation of the 1966–1970 growth, less than 2 percent per capita per year.

Economic Growth, 1971–1975

In 1971 the economy seemed to continue the slow trend of the previous five years. Gross territorial product in constant prices increased by less than 3 percent. Most of the indicators, such as exports and international reserves eased up slightly, although petroleum production declined. Price increases were less than 3 percent in 1971. The modest changes do not tell the full story. Government revenues and expenditures shot up by 17 percent in 1971, resulting in an increase in national income of greater than 9 percent.[11]

In 1972 the constant price, gross national product growth continued its moderate advance: 4 percent. In 1973 it moved up to 6 percent, but in 1974 it was back down to 5 percent. No data is available for 1975, but preliminary indications are that real growth similar to that of 1974 occurred, although not in the petroleum sector which continued to decline in real terms, and that government revenues and expenditures slid downward. The constant price calculations conceal real growth, however. National income in current prices galloped along: 12 percent in 1972, 21 percent in 1973, and an unbelievable 63 percent in 1974.

The reader must note a special characteristic of current and constant price data for Venezuela. Calculations of gross territorial product for several years use the prices of a single year, thus indicating changes in the physical volume of production. In the Venezuelan case, such calculations use, therefore, a single year's price of petroleum. A change in the price of petroleum, however, is external to the economy and represents a real change in national income through its effect on government revenue. Real national income can also increase more rapidly than gross territorial product when the share of petroleum revenues going to the government increases. The difference between current and constant price calculations is more than just price increases; current price measures also indicate real changes when petroleum prices or taxes change.

In the 1970s the share of total revenue going to the government has mounted year by year, especially since 1971 when congress authorized unilateral setting of reference prices. The share in 1975 had reached 85 percent of total revenue. In addition, the price of petroleum had increased. In 1972 it went up 12 percent, in 1973 another 13 percent, in 1974 the price tripled, and in 1975 it rose another 31 percent. With 1971 as the base, 1975 prices stood at 501 percent. On the other hand, petroleum production has declined, and in 1975 was less than two-thirds of the 1970 all-time high.

The most spectacular changes in the 1970s were in government revenues and exports, reflecting the extraordinary events in the petroleum market. In 1971 exports advanced 22 percent in value, whereas the year before the increase was small. In 1972 the increase was only 12 percent, but in 1973, the year of the crisis, exports jumped a whopping 49 percent. Then the incredible quantum leap of 70 percent came in 1974. The same pattern prevailed for government revenue which increased 17 percent in 1971, slowed to only 4 percent in 1972, but jumped 28 percent in 1973. Then again the unbelievable 1974: 246 percent. International reserves paralleled revenues and exports. In 1970 international reserves were $1,023 million; in October 1975, they were $8,781, nearly half of all of the reserves of Latin America.[12]

Government expenditures paralleled revenues until 1974. In that year it became almost impossible to spend fast enough to absorb the incoming revenues, and the economy was showing a dangerous inflationary trend. Even though 1974 expenditures were 50 percent more than expenditures in 1973, the government began to set aside some of the money in the Venezuelan Investment Fund, through mid-1975. It isolated $4.5 billion from the domestic banking system and economy, recycling the money into international capital markets. The government plans to use these funds gradually over the coming years to finance major development projects.

Expenditures did increase enough to generate the first major inflationary threat in Venezuelan history. In the past, prices have risen 1 to 4 percent, not only as a result of increased prices of domestic goods, but also through price-inflated imported goods. But in 1972 and 1973 the imported goods wholesale price index rose 5 percent each year, and the national goods index by 3 percent in 1972 and 7 percent in 1973. In 1974 and 1975 the inflation worsened, with estimates ranging from 12 to 18 percent per year.

National Income and Its Composition

National income in current prices increased from Bs1.5 billion in 1936, to Bs8.6 billion in 1950, to Bs38.0 billion in 1969, to Bs53.3 billion in 1972, to Bs102.3 billion in 1974. The average annual increase was about 8 percent from 1950 to 1969, and about 23 percent in the 1970s. The most spectacular increases were from 1950 to 1957 and 1970 to 1974. In the worst year, 1960, national income declined by 2 percent; in the best year, 1974, national income increased 60 percent. In the 1935–1974 period when national income went up about 68 times, domestic prices went up about 2.5 times and population increased about 3.5 times. Per capita national income has more than doubled each decade in the aggregate in current prices, and nearly 1.7 times each decade in constant prices.

Not all of these increases in national income have shown up directly in the incomes of Venezuelans. National income consists of payments to labor for its services and to other factors, such as land, capital and entrepreneurship in the form of interest, rent and profits. All factors pay taxes which are a part of national income. Between 1950 and 1969 payments to labor varied between 52 and 61 percent of the total. The remainder consisted of payments to domestic factors of production, including taxes. In the 1970s payments to workers were from 36 to 51 percent, on a declining trend. Indeed, since 1950 the share going to labor has declined irregularly.

The declining labor share of national income must not be interpreted, however, as a loss to labor or diminishing income. The trend reveals a major structural change in the economy. Net petroleum income does not appear in national income; it is a payment to foreign factors of production. However, taxes on that income do appear in national income as the earning of the public sector, that is, government revenue. When taxes increase faster than national income, the payments to non-labor factors tends to increase faster than national income and the labor share goes down.

In 1950 national income was Bs8,607 million and government income from petroleum was Bs1,021 million, 12 percent of the total. Between 1950 and 1960 government income from petroleum increased faster than national income and the ratio of government petroleum increased faster than national income and the ratio of government petroleum income to national income was 14 percent. By 1970 the ratio was 15 percent, and in the 1970s subsequently it continued to increase. In 1974, as a result of the petroleum price increase and swollen petroleum income, the government revenues were 38 percent of national income.

The trend of the 1950s and 1960s was unequivocal in the 1970s. The labor-non-labor share in 1970 was 50-50; but calculated in three parts, labor, non-labor and government, the shares for the first two were 49 and 19, and 36 percent went to the government. By 1974 labor received 36 percent and non-labor 64 percent; but of the latter, the government received 85 percent, whereas in 1970 it had received only 66 percent. With nationalization at the end of 1975, the previous payments to foreign factors will go to the government and become a part of national income.[13]

Gross Territorial Product

Gross territorial product, the sum of everything produced for final use at market prices, was Bs11,826 million in 1950. It was 37 percent larger than national income. The difference was 10 percent depreciation, 14 percent payments to foreign factors of production, and 13 percent for indirect taxes minus subsidies. By 1969 gross territorial product exceeded national income by only 28 percent, of which depreciation was 12 percent, foreign factor payments 10 percent, and indirect taxes minus subsidies 6 percent. The same trends prevailed in the 1970s: less for foreign factors and indirect taxes, more for depreciation, and a narrowing gap between national income and gross territorial product.

The increase in depreciation is an accompaniment of the high rate of capital formation, and the relative decline in foreign payments and indirect taxes reflect the growing emphasis on income taxes in the petroleum sector. The highest payment to the oil companies came in 1965 and 1968 before the 1973–1974 oil boom which lifted profits slightly above previous highs. This means that foreign factors have captured virtually none of the growth of the economy during the last decade. The declining proportion of indirect taxes implies that foreign petroleum clients are bearing an increasing share of the tax burden and low income Venezuelans relatively less.

From 1950 to 1969 gross territorial product, in current prices, increased 296 percent. As in the case of national income, growth was most rapid in the 1950–1957 period, at 102 percent. In the decade of the 1950s it grew by 120 percent. On the other hand, in the 1960s this measure grew only 83 percent. Since non-national income components of gross territorial product grew less rapidly than did national income, gross territorial product moved up more slowly than national income. In the last five years of the decade of the 1960s gross territorial product grew only 7 percent per year, but in the subsequent five years, 1970 to 1974, growth was 24 percent per year—65 percent in 1974 alone.

On a per capita basis, gross territorial product grew rapidly in the first eight years of the decade of the 1950s and then stagnated through 1961, advancing 51 percent between 1950 and 1957. But 1962 was less than 5 percent above the 1957 level. From 1962–1969 per capita product grew 24 percent. For the two decades, 1950–1969, gross territorial product per person advanced 96 percent. In the next five years, 1970–1974, it grew 226 percent. In the 25-year span gross territorial product per capita increased about 4.5 times.

Real rates of growth—the rates of increase of gross territorial product in 1957 prices for 1950–1969, and 1968 prices for 1970–1074—showed the same wide variations and trends that growth rates in current prices did. The principal differences arise from the mild price increases, tending to accelerate toward the end of the period, and the distortion introduced by a constant petroleum price. The amplitude of variation in growth rates diminishes when calculated in constant prices. For the entire period, 1950–1974, constant price gross territorial product grew about 297 percent.

Overall growth from 1950 to 1969 was 252 percent, about 7 percent per year. The best years were 1950–1957, an annual average of 9 percent, whereas 1958–1961, which corresponds generally with the recession, registered less than 5 percent per year growth. From 1962 to 1965 growth was stronger, 8 percent per year, but the final four years of the decade saw the poorest performance, averaging 4 percent per year. From 1970 to 1974, gross territorial product, figured in 1968 prices, also grew slowly: 18 percent in the five years, less than 4 percent per year.

Slow growth of physical output—even declining production in the case of petroleum—and rapid growth of national income after accounting for the effects of domestic inflation means that Venezuela received more income from its petroleum resources in the 1970s than in the past. The price increases of 1973–1975, however, are unique. It was more of a five-fold revaluation of the resource and is unlikely to be repeated. More modest changes in petroleum prices will result in a closer parallel between income and product.

Gross Territorial Product by Sector

The price of petroleum makes it necessary to examine two structures of production, one the physical structure as revealed by constant price measures of production, and the other, the current price measure, which reveals the financial connection between petroleum and public revenues and expenditures. Petroleum has been a declining industry since 1957 when it contributed, in the prices of that year, 30 percent of gross territorial product. In 1974, with about the same physical production but in an economy which had grown in real terms nearly 2.5 times, petroleum contributed 47 percent of total product, in the prices of that year. If 1974 petroleum output had been sold at 1957 prices and the 1974 gross territorial product recalculated on that basis, petroleum would have contributed only not quite 10 percent of total product.

Based on current prices, the share of petroleum in the economy increased from 1950 through 1957 and then declined until 1971 when it began to increase again, followed by 1973 and 1974 when the change was spectacular. The growth of other sectors of the economy displays opposite trends. The behavior of petroleum, however, does not obscure the increasing relative importance of manufacturing and of services. The relative stability of agriculture is also clear.

Variations in the petroleum share demonstrate the continued dependence and vulnerability of the economy. From 1950 through 1957 petroleum was the leading sector and imparted rapid growth to the rest of the economy. From 1958 through 1971 petroleum lagged behind the economy, which grew, under the impulse of domestic sectors only, much more slowly. Then, beginning in 1972, petroleum became the leading sector again, and real growth was rapid. The petroleum leading sector of the 1950s, however, differs from the 1970s petroleum leading sector. In the former case both output and price were expanding; in the latter, output was contracting and price going up.

The constant price measures reveal structural changes with clarity. Petroleum is a declining industry. Agriculture is stable but with a tendency toward decline. Manufacturing and services, which includes much of the government expenditures, are of increasing importance in real terms. Primary production is losing ground to secondary and tertiary production. In the 1950–1969 accounts, based on 1957 prices, the petroleum share declined from 30 to 21 percent. Agriculture declined slightly. Manufacturing increased from 10 to 13 percent. Commerce displayed no clear trend but services increased from 26 to 34 percent.

Using 1968 prices, the constant price calculation for the 1970s shows a continuation of these trends. Agriculture, however, declined slightly. Petroleum continued its downward trend. In 1974 petroleum was contributing less than 13 percent to gross territorial product, and manufacturing, which carried forward its upward trend, was contributing more than 16 percent of total production. Industrial production, which includes electricity and construction as well as manufacturing, accounted for 23 percent of total production. The share of services also increased.

Venzuela departs from the pattern of structural change experienced by most already-developed countries, a pattern in which development progresses from primary to secondary to tertiary sectors, in that order of rapid growth. In the Venezuelan case, the primary sector has indeed declined, but the shares of both the secondary and tertiary sectors have increased. Some industrialization has transpired, but the service sector did not await industrial development; it increased its share along with that of industry.

The nature of the Venezuelan primary sector and disposition of its surplus explains this apparent paradox. In the growth of most developed countries the primary sector was agriculture. Its surplus was invested in industry. When industry had developed, the service sector began to grow to support industry and the more service oriented nature of consumption. In the Venezuelan case, the primary sector is petroleum. Foreigners drained a part of its surplus abroad and the government captured some of the surplus and invested it in services, social programs and building infrastructure. Thus, services and industry develop together rather than in sequence.

In the 1970s Venezuela was structurally more an industrial country and less a primary producing country than it was in 1950. From this it would appear that the degree of dependence is less than it was in the past. The prosperity of the 1970s, however, brought to the economy by high petroleum prices, demonstrates that petroleum is still the vital center of the economy and that progress toward diversification has attenuated but not removed the vulnerability of the economy to outside forces.

Petroleum-Government-Total Product Nexus

The continuing dependence of Venezuela on petroleum is a function not only of the size of the petroleum sector in relation to total output, but also to the interrelationships among petroleum production and exports, government revenue and expenditures, and total product. Over the past four decades these interrelationships have remained close, and in some respects even heightened. While the petroleum industry has been an enclave, employing huge amounts of imported capital equipment and small amounts of labor, and sells its product abroad, it touches the economy in a strategic place: the government.

Beginning in the mid-1940s the government began to take more than 50 percent of the net revenues of the oil companies. Through the dictatorship it remained slightly above 50 percent, but in 1958 it jumped to 65 percent. Through the 1960s the government take was slightly above two-thirds. Then in 1969 it pushed above 70 percent for the first time. In 1970 and 1971 it was a little short of 80 percent; in 1972 and in 1974, 82 percent. The year of the nationalization, the government took 85 percent of oil revenues.

On the other side of the coin, the bulk of government revenues comes from petroleum. It was not always so; in 1930 only 22 percent of government revenue originated in oil; in 1940 it was 30 percent and 1950, 47 percent. By 1964 it had edged up to two-thirds and has remained at that figure since. The oil companies are the most important income tax payers, between two-thirds and three-fourths in the 1960s and 1970s. Thus, while the petroleum sector may not be an overwhelming share of the economy it is still the determinant force in the level of government activity. Since government expenditures have been running 25 to 30 percent of national income (and the government occupies a strategic place in the economy) the petroleum sector, through the government, heavily influences the level and trend of all economic activity.

Petroleum, government and the economy move parallel to one another. With few exceptions the rates of growth change in the same direction and are strong or weak together. The growth rates of petroleum exports and gross territorial product declined from 1950 to 1953, accelerated to 1957, collapsed in 1958 and recovered in 1962, followed by more modest growth rates from 1963 to 1970, and then acceleration once again.

Before 1958 the growth rate of petroleum exports was greater than that of gross territorial product. In 1958 to 1971, however, the economy consistently grew more rapidly than did petroleum exports. Beginning in 1972 petroleum exports increased once again to a faster rate than that of the economy. The lesson is clear: when the petroleum market is good, petroleum pulls the economy along rapidly, and when it is bad the economy grows slowly but still with sufficient vitality to contribute increasing per capita incomes.

Investment and Savings

Between 1950 and 1975 Venezuela set aside 20 percent of gross territorial product to maintain and increase its capital stock, a little less than one-third for depreciation and more than two-thirds for new capital. In the 1950s and the 1970s the investment rates were above the average, and for the 1960s they were below the average. The sharp break in 1960 resulted from the collapse of private investment in the recession and the decision of the oil companies to decelerate investment in Venezuela because of its aggressive policies. The private economy recovered in the course of the 1960s but the oil companies disinvested, by slowing new investment and taking heavy depreciation to avoid taxes.

Capital formation by the public sector has been a significant proportion of the total. In the 1950s the government provided about 45 percent of the total, but this declined to 35 percent in the 1960s. In the 1970s the share increased slightly, and with the petroleum bonanza in 1974 the government has earmarked substantial investment expenditures. About one-half of government investment is in productive activities, such as steel mills and petrochemical plants. The remainder goes to transportation and other social and economic infrastructure.

The sectoral shares of investment experienced great variation. Investment in mining and petroleum bouyed up net investment in the 1950s. Out of the reduced amount of investment of the 1960s, capital formation in agriculture moved ahead strongly, supported by the government. In the late 1960s and the 1970s, investment in manufacturing picked up and investment in petroleum resumed at a more moderate rate. Foreigners concentrated their investment in petroleum, three-fourths of foreign investment and industry. The greatest growth in foreign investment since 1960 has been in industry.

Since 1950 Venezuela has saved enough to finance all of its own capital formation and to invest abroad. An unknown amount of Venezuelan capital is in bank accounts and savings balances in New York, Miami and Europe. On the other hand, a net capital inflow financed a part of capital formation within the country. The government is one of the most important savers, through large Treasury balances and high and rising international reserves, augmented in 1974 by the Venezuelan Investment Fund designed specifically to promote future development projects. Savings data are not adequate but deposit data suggest that Venezuelans do save personally also in large amounts at home, enough to finance even large domestic investment and replace foreign investment.[14]

Distribution of Income

Even though Venezuela has a high average income, its distribution is highly unequal, not only among income classes, but also among regions and between urban and rural sectors. In addition, no evidence suggests that money income distribution has improved in the last twenty-five years. The statistical evidence is not complete, and serious measurement problems exist, but studies indicate that income distribution in the 1970s is about the same as it was in the 1950s, although real income distribution may have improved.

The Shoup taxation mission found that in 1957 Caracas had the highest concentration of income in the country. The average income per worker was Bs14,000–Bs15,000 a year, whereas in other towns of more than 20,000 it was Bs9,000. In rural areas it was Bs1,500, one-tenth that of Caracas. City size correlated positively with income. Cities over 20,000, including Caracas, earned 72 percent of total income and rural areas earned only 9 percent. But two-fifths of the population lived in rural areas.

The 43 percent of incomes below Bs4,800 a year earned only 7 percent of income. On the other hand, the 5 percent who received more than Bs24,000 a year earned 26 percent of all income. More than one-half of incomes earned were less than Bs6,000, and this group earned 21 percent of total income. In rural areas 94 percent of incomes were less than Bs2,400, and incomes of this amount and below were only 9 percent of total income.[15]

Other studies confirm this highly unequal distribution pattern. A study by the Central Office of Coordination and Planning indicates that in 1972, of the families which earned more than Bs100 per month, 97 percent lived in the cities. Indeed, 94 percent of the families earning incomes of Bs1,500 to Bs2,000 a year lived in cities, and those earning above that level nearly all lived in cities. Urban areas accounted for 86 percent of all earned income. But 13 percent of incomes below Bs100 per month was earned by city families and 87 percent by rural families. Another study indicates that in 1970 the lowest 30 percent of people received only 6 percent of total income, whereas the top 10 percent received 40 percent of income. Other studies in the 1960s do not show an equalization of income.[16]

A United Nations study places Venezuela among the countries with the greatest inequality in the early 1970s. The top 20 percent of income earners earned 65 percent of income, whereas the lowest 40 percent of income earners obtained only 8 percent of income. Among all countries with per capita incomes of $750 or more, Venezuela shares the dubious distinction of "high inequality" with Finland and France, both of which have more equal distribution than has Venezuela. The UN estimates that while gross national product grew about 6.5 percent from 1962 to 1970, the growth rate of incomes of the lowest 40 percent was only about 3.5 percent.[17]

Government expenditures employed specifically for socio-economic purposes since 1958 do not, however, show up in money income statistics. Educational opportunities, low-cost housing, sewers, enlarged social security, subsidized employment, health facilities, roads and other elements of social and economic infrastructure enhance real income. These kinds of activities have been a major focus of the government in the 1960s and 1970s. The relative decline of indirect taxes and increasing reliance on progressive incomes taxes tended to benefit more the lower income groups. It is not possible to measure this kind of real income distribution, but is has been of some significance.

Situation in 1975

By 1975 Venezuela had reached notable heights of prosperity and had a rapidly growing economy. Recent governments have addressed, and to some extent ameliorated, each of the four deficiencies cited earlier: dependence on petroleum, sensitivity to international markets, socio-economic distortions, and absence of regional integration. To be sure, dependence on petroleum continued, but the industry was in government, not private hands. Large reserves and the investment fund, as well as the growth of domestic agriculture and industry, provided a cushion against the vagaries of international markets.

By 1975 the spatial integration of the country was nearly complete. Venezuela had one of the most extensive and modern road networks in Latin America and virtually the entire country enjoyed an effective transportation and communication system, including extensive air routes. The government had a policy of regional economic balance, pushed regional economic planning, and was accelerating agricultural development. The electrification of the entire country was almost complete.

The most notable improvement was in the social system. The government has poured vast resources into education, housing and social infrastructure. Education at all levels has expanded greatly and Venezuela is foremost in Latin America in its support of education. Illiteracy has declined and adult education has made great strides. Twice the proportion of people receive protection from social security than in the 1950s.

Caloric and protein intake has increased substantially. Labor reforms better protect workers' rights, and working conditions have improved. Real income distribution is more equal, and the poor pay fewer taxes. While many deficiencies remain, the improvement has been substantial.

The economy, of course, remains vulnerable to the international petroleum market. Venezuela now, however, has more control than it did, including a publicly owned petroleum sector. It does not control price but through participation in the Organization of Petroleum Exporting Countries it has a voice in price decisions. In other areas, Venezuela is less sensitive to international influence. Agricultural progress has reduced dependence on imported food. Domestic industry supplies an increasing proportion of final consumption goods, and a small but growing heavy industry also relieves dependence on imports.

By 1975 Venezuela presented a more balanced posture in most respects, and the performance of the economy and public policy has reduced some of the earlier paradoxes and extremes. The institutional framework is more modern and complete. Labor organizations are robust and new legislation has raised the effectiveness of the Central Bank, the commercial banking system, and other financial institutions. The capital market is in process of reform. The public administration, despite its large size, is reasonably honest and no more inefficient than what prevails in most countries. A planning organization coordinates development activities and provides a guide and standard of performance for both the public and private sector. The government has a development policy and the financial and physical resources to implement it, although on occasion political matters are paramount.[18]

Despite these many improvements Venezuela still, in 1975, has not taken off in the Rostovian parlance. It had not reached the point where its growth was regular and sustained, generated from within itself and not vulnerable to even catastrophic reverses. Recession followed prosperity in an earlier period, succeeded by slow growth and once again prosperity, all dictated by external factors. Growth and structural change have altered the economy in fundamental ways and brought it to the brink of modernization, conferring upon Venezuelans a greater ability and willingness to control more of their economic environment. But the burden and blessing of petroleum still dominates.

NOTES

1. Banco Central de Venezuela, *Ingreso Nacional de Venezuela*, Caracas: Editorial Relámpago, 1949, pp. 93–94.
2. Price and government income data are from various editions of the *Anuario*

Estadístico. The income and product data for 1950–1969 are from an internally consistent set of accounts maintained by the Banco Central de Venezuela and published in its annual *Informes Económicos* and in *La Economía Venezolana en los Ultimos Treinta Años*, Caracas: Banco Central de Venezuela, 1971. These accounts are in current and 1957 prices. Beginning in 1968 the Banco Central changed both the system of accounts and the price base, the latter to 1968. The data for 1968–1974 are from the *Informes Económicos*.

3. *La Economía Venezolana en los Ultimos Treinta Años*, 1971, records a wide variety of economic data for 1950–1969, including national accounts, price, monetary, tax and expenditure, production and employment data.

4. "Economic Developments in Venezuela in the 1950s," *Economic Bulletin for Latin America*, Vol. V, No. 1, March 1960, pp. 21–61, assesses in some detail the position of the Venezuelan economy in 1950. The Report of the International Bank for Reconstruction and Development, *The Economic Development of Venezuela*, Baltimore: Johns Hopkins Press, 1961, also covers this period.

5. The growth rates in this and the following two sections are in constant 1957 prices unless otherwise noted.

6. Petroleum sector data are from *Petróleo y Otros Datos Estadísticos* published annually by the Ministry of Mines and Hydrocarbons.

7. *Revista Trimestral de Economía y Ciencias Sociales*, No. 1, September 1958, p. 18.

8. Carl S. Shoup and associates. *The Fiscal System of Venezuela*, 1959, pp. 21–86.

9. *Economic Bulletin for Latin America*, Vol. V, No. 1, March 1960, p. 21.

10. Ernesto Peltzer. *Ensayos sobre Economía*, 1965, p. 357 ff, covers these crisis years.

11. Income and product data for 1968–1974 employ a different methodology, different categrories of estimates, and a 1968 price base. The estimates appear annually in the *Informe Económico* of the Banco Central de Venezuela.

12. *International Financial Statistics*, Vol. XXVIV, No. 1, January 1976, pp. 406–409.

13. María Luz B. de Sanchez-Covisa. "La Estructura del Poder Económico en Venezuela," *Resumen*, 14 December 1975, pp. 14–17.

14. Organization of American States, *El Esfuerzo Interno y las Necesidades de Financimiento Externo para el Desarrollo de Venezuela*, 1972, pp. 43–131.

15. Carl S. Shoup and associates. *The Fiscal System of Venezuela*, 1959, pp. 21–24.

16. *Primera Encuesta Nacional de Ingresos y Gastos Familiares en Venezuela*, Oficina Central de Coordinación y Planificación, Caracas, 1964; *Estudio del Mercado de. Vivienda, MERIAVI-70*, Caracas: Banco Nacional de Ahorros y Préstamos, 1971; *Estilos de Desarrollo*, Caracas: Centro de Estudios de Desarrollo, 1971 *Encuesta de Hogares por Muestreo*, Dirección General de Estadística y Censos Nacionales, 1971.

17. S. Jain and A. Tiemann. *Size Distribution of Income: A Compilation of Data,* Development Research Center Discussion Paper No. 4, Washington: International Bank for Reconstruction and Development, 1973, cited in *Finance and Development*, Vol. 11, No. 3, Spring 1974, pp. 1–8.

18. The development of the Venezuelan economy in the last few years is

reviewed in the weekly magazine *Resumen*, in the monthly *Ve Venezuela*, and in *Venezuela Now*, a publication of Venezuela's United Nations office. The most up-to-date statistical information is in the monthly *Boletín Mensual* of the Banco Central de Venezuela.

Chapter VII

The Public Sector

Over the past four decades the public sector has grown in size and complexity, and for two decades has dominated economic life in Venezuela. It has been the chief agent of growth and transformation, and since 1959 it has directed a long-term development and socio-economic improvement program. Its function is to convert the bounty of the country's resources (paid primarily by foreigners through the purchase of petroleum) into income, institutions and assets which will produce a growing, stable and diversified economy.

With some backsliding not uncommon to governments, the Venezuelan public administration has become more efficient and more effective. No evidence of large-scale peculation has come to light and public officials have become accustomed to handling large sums of money. Budgeting by program, strict accountability and auditing, and coordination and planning within all units of government and among units has improved performance and the quality of decision-making and execution of public activities.

The newness of many governmental organizations, the inexperience of personnel, the complexity of public activities, and politics reduce the efficacy of the public sector. To perform new functions the government had to create new institutions and staff them with people without training and experience. As the public sector grew, it became more complex, functions overlapped, and planning and coordination became more difficult. Political parties intruded in bureaucratic processes; each party staffed important positions with its own people, who were not necessarily most suited. Government often limped along without long-established traditions, an adequate civil service, and a secure professional staff.

Structure of the Public Sector

Despite changes in government and new constitutions (the latest in 1961) the economic functions of government have remained much the same. The difference has been in the policies of government and the manner and degree of fulfillment of the functions. The government in 1959 accepted a number of responsibilities not recognized by previous governments, wrote them into the new constitution, revised the means by which the functions were performed, improved the public administration, and coordinated it more effectively.

The new constitution retained the basic form of government, a nominal federation with a strong central government divided into three branches: executive, legislative and judicial. Before and after 1961 the executive branch was the real seat of power, but after 1961 the congress had and used specific powers, most notably as the forum for debating and validating policy and for the control of the pursestrings. The constitution gave responsibility for social and economic welfare, maintenance of full employment, and regulation of economic activity to the government, with a long list of general economic charges.[1]

The executive and congress converted the constitutional charges into specific legislation and created institutions to carry out the purposes of the law. For example, congress spelled out in detail, in 1966, the constitutional labor provisions which guaranteed the right to organize and bargain collectively, the right to strike, and the 48-hour work week—and assigned the Ministry of Labor the task of administering the legislation. The constitution guaranteed indemnification upon expropriation, and later legislation specified the nature of the indemnity and created organizations to manage nationalized property.

The constitution and subsequent legislation assures property rights and private enterprise, subject to the government's responsibility to direct and promote the economy, build industry, and conserve natural resources. The constitution specifically charges the lower house of congress with approval, rejection, or alteration of the budget submitted by the president. An independent Comptroller General audits income, expenditures and assets; and the planning office, a part of the executive but responsible also to congress, coordinates governmental activities to promote economic development.

The executive consists of the President and his cabinet ministers, the supreme policy making body. Congress may censure but not remove a minister. Each ministry has assigned to it other governmental bodies so that the entire government has representation in the cabinet. Five types of governmental units constitute the executive branch: the central government, the regional (state) governments, the municipal governments, administrative entities and state enterprises.

The thirteen ministries which constitute the central government administered 29 percent of the total government budget in 1974. States and municipalities, which derive most of their funds from the national budget, were responsible for 6 percent and 3 percent respectively of government spending, embracing the administration of 20 states, two federal territories, and 147 municipalities. Seventy administrative organizations—ranging from universities, to the social security system, and the national children's council—have charge of 13 percent of the budget. Fifty-six state enterprises—of which 27 are financial institutions and 29 are industrial, transport, or service enterprises—spend 48 percent of the budget.[2]

Control of spending has moved strikingly in favor of administrative agencies and state enterprises. The latter, which generate and spend some of their own income through sales of goods and services, depend on the budget to cover asset acquisition and deficits when necessary. In 1960 state enterprises accounted for only 23 percent of the budget, compared to the present 48 percent, which will swell enormously with the addition of Petróleos de Venezuela, the nationalized petroleum holding company. Administrative agencies have also doubled their share of the budget, while the share controlled by the central, state and municipal governments in 1975 is only a little more than one-half what it was in 1960. Note also that special funds, notably the Venezuelan Investment Fund, set up to hold the unabsorbed petroleum revenues, are not in the above classification since when spent, these funds will go to other organizations.

The budget is the principal instrument through which the public sector influences the economy. All units of government submit their budgets (nominally program budgets) to the ministry to which they are attached. The ministry scrutinizes and in some cases amends these budgets, adds its own budget and submits the consolidated budget to the council of ministers, which refers it to the Central Office for Coordination and Planning. The coordination accomplished, the president submits the budget to congress, along with revenue estimates. Congress may modify the budget as it pleases and may unbalance, initially or through supplemental appropriations, the balanced budget submitted by the President.

The proliferation of government organizations and enterprises, subject to only loose controls by the ministers and the planning office, implies that it is extraordinarily difficult to employ the budget as a development measure. Each of the 140 organs of government, as well as some sub-units of these units, has and uses political influence within the government and in congress; each, of course, tends to identify national development with its own development. In addition, the link between revenues and petroleum exports weakens the budget as a development instrument.

Even so, in a period of rising revenues, empire-building, waste and imcomplete controls have performed a development function. As the economy has grown, many new functions have become necessary. Many enterprising individuals have gravitated toward the public sector to undertake these functions, serving thereby not only their own interests, but also those of the public and the economy. Competition within the public sector tends to restrain free-wheeling bureaucrats, even more so than competition regulates the private economy.

Government Income

In 1935 government income was Bs206 million; in 1974 it was Bs42,595 million, the latter figure not including government enterprise income. Even eliminating the doubling of prices leaves government real income 100 times higher than 40 years ago. This startling increase is somewhat deceptive, however, since much of it occurred in 1973 and 1974. Income in 1973 was Bs16,054 million, which signifies a jump of 265 percent in 1974 alone, and the increase for 1973 over 1972 was 32 percent. Since 1950 government income has gone up every year except 1958 and 1962—the latter a recession year and the former a result of income having been pushed up the two preceding years by concession payments.[3]

Government income has been highly unstable. In 1936 it went up 7 percent, the next year 41 percent. In 1941–1949 income rose 6.8 times, but in 1961–1969 it went up a mere 37 percent. In 1951–1959 the increase was 2.5 times. Petroleum accounts for the volatility of government income since it contributes about two-thirds of the total. In the petroleum prosperity of the 1940s, 1950s and 1970s, government income rose rapidly; in the stagnation of the 1960s, increases were modest.

The income of the public sector comes from income taxes, royalties, national import duties and indirect taxes, state and municipal indirect taxes, and the sale of goods and services. State and municipal taxes are minor; the principal income of these governments comes from central government tax and royalty income. National indirect taxes have been a declining proportion of income and income taxes a rising proportion. With the increase in government enterprise, income from the sale of goods and services has increased relative to income, although these units continue to depend on central government income to cover deficits.

Government income has grown faster than the economy. In 1950 government income was 17 percent of gross territorial product. By 1960 it was 26 percent. The slowdown in income reduced the ratio during the 1960s. The growing importance of administrative units and government enterprises has again, in the 1970s, pushed up the quantitative role of the public sector. General government plus enterprise income was 31 percent of gross territorial product in 1972, and 33 percent in 1974. The addition of the petroleum and iron ore industries to the public sector will make it overwhelmingly the most important part of the economy.

The government introduced the income tax in 1943 to raise more revenue, especially from the petroleum companies, and to make the tax system more progressive. Tax rates have been changed on several occasions, and a stand-by additional tax in 1948 required that the petroleum companies pay at least 50 percent of their net income to the government. In 1958 the portion reserved to the government increased to two-thirds. The income tax quickly became the most important single tax, exceeding import duties by 10 percent by 1950. At that time it was only one-half of royalties, primarily from the petroleum industry.

Income taxes went up nearly twice as fast as government income during the 1950s; income trebled and the income tax take increased 5.3 times between 1950 and 1959. Income taxes exceeded royalties for the first time in 1958. In the 1960s income taxes continued to grow more rapidly than all income. In 1960 the income tax was 20 percent of income; in 1969 it was 42 percent. By 1969 the income tax exceeded royalties by 30 percent. In the 1970s income taxes continued to grow more rapidly than other sources of revenue. In 1973 it was 62 percent of all revenue, and in 1974 69 percent. The petroleum sector has paid two-thirds to four-fifths of all income taxes in the last decade, the highest ratios since the petroleum boom began in 1973.

The government has changed tax rates several times and has broadened the tax base in order to include more contributors, especially Venezuelans, dipping further down the income ladder. Rates are also now more progressive. The last major reform was in 1966, and yet another was under consideration in 1975. At present the tax rates are only mildly progressive, and for individuals, exemptions and deductions are high. In addition, it is easy to evade the income tax, to report incorrectly, or not report at all. The government withholds for its own employees but most employers do not. The administrative machinery is creaky and inadequate to the enforcement task.

A broader tax base, higher rates, efficient collection and enforcement are the goals of the government in domestic taxation. It is fearful of the day when petroleum income will decline, and without something to replace it, the public sector may contract. The government wants a long lead time so that individuals and companies become accustomed to paying taxes. The political opposition, however, is fierce. The private sector feels that while petroleum can pay, it should pay, that the private sector should expand to the highest degree possible, free of burdensome taxes, while it can, and contribute only when it is necessary. The last confrontation of this nature, in 1966, resulted in a three-way compromise in which the government won at the petroleum sector's loss, with little damage to the private sector.

The royalty is a payment for the alienation of the patrimony of the nation. It is not a tax. The basic royalty in Venezuela has been 16.67 percent of the product extracted, payable in kind or cash. The pricing formula for determination of cash payment is similar to that of reference prices for income tax calculations. The growth of royalty payments has been more modest than income taxes. Petroleum royalty payments doubled between 1950 and 1957, doubled again by 1964, but increased only 40 percent between 1964 and 1973. Royalties increased 2.8 times in 1974. Royalties are due on other minerals (most notably iron ore), but both the amounts and the rate of growth have been small.

The public sector has been critically dependent on petroleum income taxes and royalties. With nationalization, petroleum becomes a part of the public sector. Since the government will still need a large part of the net revenues of Petróleos de Venezuela, the petroleum sector will continue to be the principal fiscal support for the entire public sector. On the other hand, as the largest single element in the public sector, Petroven will have some say in the disposition of its surplus. Costly exploratory work in the Orinoco Tar Belt, expensive technological development, and efforts to penetrate the international oil marketing and distribution system, as well as net investment to maintain and expand the present industry will make large demands on its income and make obvious the need for a more balanced tax system.

Indirect taxes, once the bulwark of the fiscal system, have declined steadily in importance. In 1936 indirect taxes, mainly import duties plus modest royalties, were all of government income. In 1950 indirect taxes were still 42 percent of government income. In the 1950s indirect taxes continued to lose ground, constituting 26 percent of government income in 1959. The abrupt increase in 1961–1963 was in reality profits on exchange transactions. The decline in indirect taxes continued in the 1960s, and by 1973 they were only 11 percent of government income. The large increase in income taxes and royalties in 1974 reduced indirect taxes to 7 percent of the total.

Before 1957 the government did not borrow and had almost no debt. In the ensuing period of political uncertainty and recession, however, the government did borrow, 19 and 17 percent of 1960 and 1961 income, respectively. Net borrowing then declined, only to resume again in 1969, but this time on income earning projects. Other income consisted of concession payments in 1956 and 1957, some 22 percent of income in those two years, and later other minor income from coinage and revaluation on an irregular basis.

Public sector income is larger than central government income. The two other levels of governments, states and municipalities, generate a small amount of income, all from indirect taxes. The amount of money in the case of state governments is minute, and the states subsist on transfers from the central government. Municipal governments have some indirect taxes but even in recent years these taxes are only one-fourth of national indirect taxes and 3 percent of central government income. Municipalities derive much of their income from transfers from the central and state governments.

The two remaining sources of public sector income are administrative units and government enterprises. Their incomes have grown rapidly, especially in the late 1960s and 1970s, not only on the basis of their increasing sales of goods and services, but also by increasing contributions from the central government. These organizations represent the branching out of the government into new areas of service and productive activity.

Carl Shoup in his study of the Venezuelan fiscal system in 1958 observed four striking characteristics: (1) nearly all revenues flow to the national government, (2) the major part of revenues come from petroleum companies, (3) most non-petroleum revenues come from indirect taxes, and (4) the tax system was erratically and mildly progressive. The first two continue to be true. But in the last two decades the non-petroleum income taxes have grown faster than indirect taxes, and the tax system is more progressive. In addition, public sector income increasingly includes non-tax income.[4]

Expenditures

How the government has spent its money is a record of its priorities. This is especially true when the government uses debt sparingly and endeavors, year by year, to spend what comes in. In the Venezuelan case, expenditures indicate that military governments emphasized the economic, administrative, financial and public works sectors; whereas civilian governments since 1958 have emphasized social and infrastructure sectors, including education and health.

Most central government expenditures go through the thirteen ministries, including the allocations of the central governments to administrative units and government enterprises. Independent state and municipal expenditures made from their own revenues, and the earned income of administrative agencies and government enterprises, are excluded from ministerial expenditures. The growth of these expenditures closely parallels government income, with some adjustments for deficits and surpluses. All central government allocations to state and municipal governments are in the budget of the Ministry of Interior Relations.

The Ministry of Public Works, in charge of all public construction, dominated ministerial expenditures through the 1950s. In 1950 this Ministry spent 36 percent of total expenditures, the same ratio as in 1957. In the 1960s Public Works began to lose ground; in 1961 it was down to 19 percent, but it began to creep up again in the late 1960s. In 1969 Public Works spent 25 percent of total, but again in the 1970s it was down to 16 and 17 percent in 1972 and 1973 respectively. The Ministry of Public Works was quantitatively the largest ministry, surpassed only by the Treasury in 1962, by Interior Relations in 1972 and 1973, and by Education in 1972 and 1973. In 1974 Public Works was once again the biggest spender. Its growth from 1950 to 1973, was only a little more than half that of expenditures as a whole.

The Ministry of Education has been the fastest growing ministry, and finally in 1972, caught up with and surpassed Public Works for two years. It started out in 1950 with a budgetary allocation only 17 percent of that of Public Works and 6 percent of the total budget. Even in 1957, the last year of the dictator, it had only 4 percent of the budget, but in the 1960s and 1970s its share grew by leaps and bounds; in 1973 it took 19 percent of the budget, more than all ministries combined except Interior Relations, the Treasury, Defense, Public Works and Agriculture. Even the allocaton to the Ministry of Education understates the emphasis on education, since many other ministries have educational programs, and Public Works builds school buildings.

The Ministry of Interior Relations is also a large ministry and its expenditures have grown more rapidly than total expenditures. The states and municipal governments receive their allocation through this ministry. The Defense Ministry, also large, receives a little more or less than 10 percent of the budget. The Development Ministry was growing rapidly until 1957 but has since stagnated. Many of its functions were shifted to other ministries in the 1960s. Budgetary allocations to the Ministry of Health and Social Assistance have gone up more than eight times, and to the Ministry of Agriculture more than 10 times.

The pattern is clear: Public Works has given way to Education, Agriculture, and Health and Sanitation, while Interior Relations, Treasury, and Defense have held their own or gained slightly. The three growing ministries received 18 percent of the budget in 1950, but 33 percent in 1973. Public Works has declined from 36 to 17 percent in the same period, while Interior Relations, Treasury, and Defense increased from 31 to 39 percent.

Analysis of data of functional categories of expenditures reveals the same basic trends. During the 1950s, especially in the middle of the decade, heavier expenditures occurred in the economic and financial categories, with reduced expenditures in administrative, social and cultural categories. Health, social security, labor and educational expenditures lost ground until the end of the dictatorship but then recovered rapidly in 1958–1960. The economic emphasis accelerated up to 1957–1958 and then declined in relative importance. The most notable changes after the departure of the dictator were relative declines in housing, transportation and industry, and relative increases in health and social assistance, social security and educational expenses.

The shift away from economic to social expenditures continued through the 1960s. Social expenditures claimed 26 percent of expenditures in 1960, 29 percent in 1969, and 33 percent in 1973, with the greatest emphasis in education, sanitation, housing and urban development. On the other hand, the strictly economic functions received relatively less. Whereas in 1960 expenditures in the economic categories were 40 percent of total expenditures; the share in 1969 was 28 percent. The same trend prevails in the 1970s.

In only one of the economic sectors did expenditures grow rapidly: agriculture. Between 1960 and 1973 expenditures in agriculture more than tripled while expenditures in the economic category as a whole did not quite double. In the 1970s expenditures in mining, manufacturing and construction began to increase more rapidly than expenditures in the economic category generally, at the expense of roadbuilding. Expenditures in the political, administrative and defense categories maintained a nearly constant proportion of total expenditures since 1960.

Enrique Baloyra studied ministerial expenditures in Venezuela in order to determine differences between priorities in military and democratic governments between 1938 and 1959.[5] By attaching particular year's ministerial expenditures to either the authoritatian or democratic camp, and then by summing years for each ministry for each camp, he sought a pattern of difference between the two types of regimes. He regards his results as disappointing, but did find that in education, welfare and health the democratic regimes had significantly higher shares of total expenditures. Many problems persist in this form of analysis, including assignment of year's expenditures to regimes, time lags at change-overs of regimes, and shifts of emphasis within ministerial expenditures. It does, however, tend to confirm the conclusion here that governments since 1959 have tended to use the public sector in pursuit of more socially-oriented goals than previous governments.

Current and Capital Expenditures

The division of expenditures between current and capital expenditures and among their components reveals how the public sector perceives its tasks and uses resources to address them. Personnel expenditures, for example, were only 30 percent of total expenditues in 1950 and the share declined through 1957, the share of the total falling to 15 percent in the latter year. In 1958, however, the share jumped to 21 percent and has been increasing ever since. The average allocation for 1950–1957 was 26 percent, for 1960–1969, 31 percent. Even this understates personnel expenditures, since much of the transfers to state and municipal governments goes to employ people.

Current expenditures follow a similar pattern. Between 1950 and 1957 current expenditures grew 75 percent while total expenditures went up 2.6 times, whereas from 1958 through 1969 current expenditures grew 35 percent and total expenditures by 57 percent. In the 1970s current expenditures grew almost as fast as total expenditures: 41 percent compared to 46 percent for the total. Much of the current expenditures goes to personnel, but increasing amounts go to administrative units and government enterprises, as well as state and municipal governments.

While the dictatorship, from 1950 to 1958, focused more on capital investment, such expenditures continued to be consequential later on as well. In 1960, for example, capital investments were 34 percent of government expenditures and this share crept up slowly during the decade, reaching 39 percent of the total in 1969. In the 1970s, however, capital investment by the government stopped growing and in relative terms even declined in the early 1970s. In 1973 it was only 28 percent of the total. In 1974, however, under the impact of petroleum revenues, it shot up to 60 percent, some of it in the Venezuelan Investment Fund.

The developing pattern of expenditures was partly a result of policy goals and partly a result of circumstances. The military regime before 1959 fancied it could ignore social problems and did not perceive unemployment as a serious matter. Governments since 1958, however, were by propensity and necessity more sensitive to the social problems of the people and growing unemployment, especially in the early 1960s. The personnel expenditures of the government rose rapidly and the government often acted as the employer of last resort. These governments, subject to complex political pressures, defended the government payrolls less ably than can a dictator. They tended to solve problems of unemployment by putting people to work in the bureaucracy or leaning on shovels for the burgeoning ministries. More money addressed social problems also. The budget category which had to yield was capital expenditures.

The military regime tended to keep expenditures under the close control of the central government. It was not wholly successful and indeed permitted some government enterprises and administrative units to run up large unauthorized short-term debt, especially on favorite projects of the dictator. Democratic governments subsequently loosened up, created more enterprises and organizations with tenuous relations to the central government, and provided less stringent surveillance.

The government since 1959 has justified greater allocations to state and regional governments, administrative agencies and government enterprises on grounds of efficiency. In a large government it is best that specialized functions be performed by organizations having that unique task, by relatively small and supposedly more responsive units. In addition, government enterprises permitted the institutional flexibility to engage in productive activity in a framework similar to that of private enterprise.

What may have been gained in administrative efficiency, however, was often offset by lack of effective control over expenditures. While officially all government operations, including state and municipal governments, administrative agencies and government enterprises are under the purview of the Comptroller General, and conform to the same budgeting and financial discipline as the ministries which nominally supervise them, in fact, they have considerable independence and political influence on their own. CORDIPLAN coordinates their activities and plans but does not have plenary powers over their expenditures. The government, therefore, is in the position of funding a substantial amount of activity which it does not control and which does not necessarily conform to its development program.[6]

Surpluses, Deficits and Debt

Since the foreign intervention in 1906, Venzuelan governments have avoided debt whenever possible. The large petroleum revenues have, in the last half-century made it unnecessary to incur debt. General Gómez regarded a debt-free hacienda as a prosperous hacienda, and for nearly three decades Venezuela was his hacienda and he paid off all internal and foreign debt in the early 1930s. These factors, plus long years of zero debt, have led Venezuelans, acting in official capacities, to regard borrowing as almost always bad. The presentation of balanced budgets and the need for congressional approval to incur foreign obligations are reflections of this prejudice against debt.

Venezuelan authorities overlooked the positive role of debt and the disadvantages of having no debt. As a device to regulate the level of economic activity and stabilize growth, debt (even foreign debt) can play a vital role. As an instrument for providing liquidity to the financial system, for developing capital markets, and for control over financial institutions, internal debt is critical. Debt is also a means for altering the timetable of development, permitting the early accomplishment of projects and paying for them out of income earned from their operations.

But the Venezuelan government has shunned debt, and only in the late 1960s and 1970s has it begun to employ debt positively. At the beginning of the 1950s almost no debt existed. During the 1950s a modest internal, indirect debt built up; these were obligations of administrative units and government enterprises. Because of concession payments, at the end of 1957 the Treasury surplus was Bs2,355 million and the acknowledged internal indirect debt was Bs915 million. A hidden debt existed, however, at this time. Government enterprises and administrative units had contracted to pay Bs4,578 for work performed. These were short-term notes, most of them discounted and sold to foreign banks. The dictator left obligations far beyond the Treasury balances.[7]

In 1958 it was necessary to begin to deal with the dictator's fiscal irresponsibility. In that year expenditures went up sharply, income by a lesser amount, and the largest deficit in Venezuelan history resulted. Large debt payments were made and some of the contract debt was converted to floating debt. Treasury reserves plummeted and in 1960 the government had to borrow abroad. That year the debt was nearly triple that of the 1957 debt and one-half of it was direct debt. In the three years of 1959–1961, Venezuela borrowed Bs4,083 million, the difference between expenditures and income after Treasury balances had all but disappeared.

Venezuela's handling of the dictator's financial affairs was sound financial policy but dubious development policy. The entrenched attitude toward debt was such that any debt had to be paid. Many of the obligations, however, were highly dubious, had been heavily discounted, and payment in full enriched speculators and bankers more than responsible debt holders. Bear in mind, however, that these were tenuous times for

Venezuela; capital was flowing out and the government was none too secure. Authorities felt that paying the debt would bolster Venezuela's international credit, which it probably did, but more than a year's income was lost, resources which could have developed the economy and fought the recession.

In 1961 the public debt was Bs2,778 million, of which two-thirds was direct obligation and the remainder had a government guarantee. The foreign debt was 29 percent of the total. From 1962 to 1967 there were surpluses, some debt repayment, some new debt. Treasury balances climbed. The total debt grew only 2 percent in these years but in 1967 more of it, 53 percent, was foreign debt. The floating debt disappeared. Venzuelan authorities by this time were becoming adjusted to a small debt and to observe its usefulness in monetary affairs and for development purposes.

In 1968 and 1969 deficits recurred but they were small. The debt, however, increased significantly, 61 percent in the two years. By 1970 the debt was double that of 1967, foreign debt having climbed at the same rate as domestic debt. Despite the surpluses of the 1970s, the debt continued to go up. At the end of 1974 the total debt was Bs8,651, greater than the 1970 debt by 57 percent.[8] The nationalization of the iron ore and petroleum sectors will result in an increase in debt of about Bs5,000 million.

State and Municipal Finance

The state and municipal governments are inert. They live off the central government. The national government limits their ability to tax, and the amounts are small. In the case of state governments, nearly all income comes from the central government and has increased faster than has the income of the central government. Eighty percent or more goes to current expenses which have grown faster than their investment expenditures. States have trivial amounts of debt and very small surpluses and deficits.

Even though the states are merely an expenditure channel, they do handle a significant amount of resources. In 1960 states spent 10 percent of the total budget, and this share has crept up over the years. In 1972, for example, states spent 14 percent of the total budget.

Municipal governments have fewer resources available to them and of that which they have, some comes from their own revenue sources, mainly indirect taxes. Between one-fourth and one-third of the resources of municipalities comes from transfers from the central and state governments. Its indirect taxes, principally the patent, which is a license to do business based on the size of the firm, are about one-fifth of the indirect taxes collected by the central government. As in the case of state governments, current expenditures predominate; 20 percent goes to investment, and this proportion has declined in recent years. Municipal indebtedness is small, as are surpluses and deficits. The Federal District, which includes Caracas, spends slightly more than one-half of all municipal expenditures.

Administrative Entities

Administrative entities are special government organizations not under the direct supervision of any ministry. They include the universities, regional development corporations, and cultural, technical and social organizations. Some are autonomous institutes and others are foundations. Each ministry has several administrative entities operating nominally under its care, but the entities are in fact independent.

Administrative entities handle more financial resources than state and municipal governments combined. In 1960 they had an income and expenditure of Bs944, of which 73 percent came from transfers from the central government. Their own revenues grew more rapidly than central government contributions, and by 1969 total income more than doubled, of which about one-half came from the government. In the 1970s they continued to grow rapidly, growing 42 percent between 1970 and 1973, with government incomes increasing slightly more. Some 70 to 75 percent of their expenditures are for social purposes and another 10 percent for economic purposes.

Government Enterprises

The government owns and operates financial, industrial, commercial and transportation firms. They are large and growing rapidly. In 1975 the iron ore industry joined the group, and in 1976 the petroleum industry becomes a number of companies all held by Petróleos de Venezuela. A grant from the government constituted the original patrimony of these establishments and the government makes up any deficit. Growth was rapid throughout the 1960s, and from 1969 to 1973 they expanded in expenditures 87 percent, substantially more than the 53 percent expansion of central government expenditures. Income from government enterprises was 7 percent of all government income in 1973.

The Public Sector

No acceptable measures of the efficiency of the public sector exist. Critics abound, both inside and outside the government, and political leaders, as well as the government itself, are continually striving to improve the quality and reduce the costs of operating the public administration. On-the-job training, in-house training, encouragement to take special courses, continuous pressure, budgeting by program and many other devices encourage the bureaucracy to improve productivity.

Convincing evidence exists that during the Pérez Jimémez dictatorship corruption and maladministration at public expense was widespread, reaching to the top. The democratic regimes since 1959 have endeavored to keep the government in better order with more determined and detailed accounting, and the government is now reasonably honest. The burgeoning payroll and political exigencies of a democracy have resulted, however, in an unwieldy and complex government apparatus which is seeking but has not found greater efficiency.

NOTES

1. Thomas E. Weil and others. *Area Handbook for Venezuela*, 1971, pp. 205–230, covers the governmental framework.

2. Mauricio García Araujo. "El Gasto Público Consolidado," *Resumen*, 26 October 1975, pp. 8–14.

3. Data on government income and expenditure comes from *La Economía Venezolana en los Ultimos Treinta Años*, 1971, and from the annual *Informe Económico* of the Banco Central de Venezuela, as well as the *Anuario Estadístico*.

4. Carl S. Shoup and assoicates. *The Fiscal System of Venezuela*, 1959, pp. 2–6.

5. Enrique Baloyra. "Oil Policies and Budgets in Venezuela, 1938–1969," *Latin American Research Review*, Vol. IX, No. 2, Summer 1974, pp. 28–72.

6. Fred D. Levy, Jr. *Economic Planning in Venezuela*, 1968, pp. 128–157.

7. Carl S. Shoup and associates. *The Fiscal System of Venezuela*, 1959, pp. 476–477.

8. *International Financial Statistics*, Vol. XXVIV, No. 1, January 1976, pp. 408–409.

Chapter VIII

Human Resources

Against great odds, Venezuela has developed and deployed its human resources in an increasingly proficient manner in recent decades. The rapid rate of population growth, age distribution, internal migration, lack of skill, productivity, and motivation—coupled with a dual wage structure, labor organizations and the politicization of the labor force, unwise use of capital goods, unemployment, and other factors—have all tended to impede economic growth since 1950. On the other hand, the labor force has matured, acquired skills and become more productive; the government has mounted a giant training and educational program. Labor organizations play an increasingly responsible role. Great progress has been made in the development and use of the country's human resources.

Population Characteristics and Impact

The population of Venezuela was 3,467,000 in 1935, 5,035,000 in 1950, and 10,722,000 in 1971, according to census reports. The year-end estimate for 1971 is 11,150,000, for 1972 it is 11,520,000, and for 1974 it is 11,901,000. Venezuela is the sixth most populous country in Latin America, and while population density has increased in recent years, it is still much lower than in most countries.[1]

The four principal characteristics and trends in population in the last quarter century have been: (1) rapid rate of population growth, (2) relative youth and low participation of the population in the labor force, (3) uneven geographic distribution and concentration regionally and in urban centers, and (4) massive internal migration, especially from rural to urban areas.

Three trends of population growth mark the time since colonial days. From 1800 to 1920 population grew slowly, tripling in 120 years. From 1920 to 1941 the rate of growth accelerated to more than 2.5 percent per year, and since 1941 population has grown at 3.5 percent per year, among the highest in the world. In the last few years the population growth rate has tended to decline as the birth rate has fallen and the death rate has stabilized. The birth rate per thousand was 45.3 in 1961, 41.6 in 1966, and 38.1 in 1971. The death rate per thousand was 7.3 in 1961, 6.8 in 1966, and 6.6 in 1971.[2]

The falling birth and death rates are attributable to more knowledge, public health measures, and improved socio-economic conditions. Before 1936, for example, Venezuela was among the highest in malaria incidence. The government sprayed the country with DDT and the disease disappeared. Venezuela has now joined more advanced countries with heart disease and cancer as the most common killers. Education, a relaxed church attitude, higher incomes, lesser need for children as social security, and greater sense of responsibility are bringing the birth rate down.

The population has become younger as it has grown rapidly. In 1951, 51 percent of the population was 19 years of age or under. By 1961 this proportion was 56 percent. Ten years later the same proportion was 16 years of age or younger. This age structure has resulted in a low and declining ratio of number of people in the labor force to total population.

The economically active population grew more slowly than did population, not only because of the younger population but also because young people were kept in school longer. In 1950, 34 percent of the population was in the labor force, 32 percent in 1961, and 30 percent in 1971. The burden of production has thus fallen on a declining proportion of the population, all of which continues to consume.

Not only did the young people not produce, but also the government had to divert resources from economic infrastructure and production stimuli to educational facilities, social services, housing and social infrastructure. Governments since 1959 have made extraordinary efforts to satisfy social aspirations including those of the young, at the expense of immediate growth.

The larger population, on the other hand, along with economic growth has contributed to the enlargement of the market, making possible the domestic production of goods not possible before. In addition, as these young people grow up they are experiencing a more congenial social and economic environment, with more and better educational and training opportunities. They enter the labor force with more skills and higher productivity, and therefore greater incomes, which reinforces the expansion of the market.

Migration and rapid urbanization has also burdened the economy. As in the case of the population surge, the move to the cities began in the 1920s and continues in the 1970s. In 1936 only 22 percent of the population lived in urban communities of 1,000 or more inhabitants. By 1941 the proportion was 39 percent; in 1950 it was 46 percent. The movement has continued in more recent years. By 1961 the cities accounted for two-thirds of the population, and in 1971 for three-fourths. The ten largest cities in 1950 had a population of 1,407,000, 28 percent of the population. By 1961 the proportion living in the 10 largest cities was 36 percent, and in 1969 it was 41 percent. In 1926 only 4 percent of the population lived in Caracas, but in 1971 more than 20 percent lived there. Caracas has grown twice as fast as the country, but even it was not the fastest growing city.[3]

New housing, education facilities, social and health services, transport networks, and other infrastructure were necessary to absorb the people moving into new areas. Sufficient classrooms exist for all school-age children but critical deficiencies exist in some areas with unused capacity in others. In addition, much of the population movement took place with only limited employment opportunties. People often arrived in cities without resources and faced long periods of unemployment or secured employment by displacing others.

The difference in real income explains the migration to the cities. The cities offer higher incomes, more economic and social opportunities, a sense of participation, and more congenial surroundings. Agriculture offers poorer prospects, lower income, and a mean living. The money income is significant and even if unemployed, an individual may be as well off in the city, living with friends or relatives.

Urbanization has reinforced the unequal distribution of population among regions. The greatest concentration of people is the capital region, which in 1971 had 25 percent of the population but less than one percent of the land, a population density 2.5 times the national average. The proportion in the capital region has been growing; it was only 20 percent in 1950. At the other extreme is the Guayana and the southern regions which in 1971 had 57 percent of the land but only 7 percent of the population. The industrial complex and new city in the Guayana has attracted population, but the southern, as well as the east-central and northwest regions have lost population relatively. The Andes, with 14 percent of the population in 1971, has 9 percent of the land but continues to lose population to Zulia and the capital region. The greatest population concentration is the capital and central regions which had 33 percent of the population in 1950, 40 percent in 1971.[4]

The foreign-born are a small proportion of the population, only about 8 percent in 1969. Most are white and came from Spain, Italy, Portugal and the United States. Most are naturalized citizens. Between 1945 and 1959, 397,000 foreigners entered the country, but since 1959 when the government tightened immigration laws relatively few have arrived. In 1975 the government decided to encourage immigration again, especially of professionals and technicians. In many cases foreigners and first-generation Venezuelans are in the upper strata of professional, managerial and technical occupations. In addition to the official foreigners, Venezuela is also reluctant host to more than 500,000 Colombians who crossed the border illegally and who live and work in Venezuela.

Labor Force

Remarkable changes occurred in the sectorial and occupational structure of the labor force. The most important developments were the absolute and relative decline in agriculture and petroleum, and relative increases in industry, commerce and services. In 1950 agricultural employment was 44 percent of the labor force. In the early 1970s the sector had only 25 percent (including the 350,000 in the campesino sector,) and the number of agricultural works was stable. Petroleum and mining employ few workers. In 1950 the petroleum labor force was 45,000, declining in the 1960s as exploration and expansion of fields—the only parts of the industry with large labor requirements—slowed down. In 1972 the petroleum labor force was only 23,000.

The industrial sectors expanded their share of the labor force from 16 percent in 1950, to 27 percent in 1971, reflecting industrialization and more labor intensive manufacturing in the 1960s and 1970s. In the 1950s construction boomed and employment increased, but in 1959 the industry began to contract, not regaining its former position until the mid-1960s. In recent years construction employment has increased. Electric power's small labor force has expanded rapidly. Construction doubled its share of employment between 1950 and 1971, using 19 percent of the labor force in the latter year. Transportation, communications and services employment all expanded. In 1971 services alone employed one-third of the labor force.

Employment of professionals, technicians, managers and administrators has grown more than twice as fast as the labor force since 1950. The share of office workers increased significantly, as did that of white collar workers as a whole. On the other hand, the number of people who classified themselves as agriculturalists dropped from one-third to one-fifth. The shifts in occupational specialties show clearly the trend away from the primary to the secondary and tertiary sectors. Women also play a greater role. In 1970 women were 21 percent of the labor force, compared to 12 percent in 1960. Nearly one-half of the service labor force are women, many of whom are in domestic service.[5]

Ouput per Worker

The most striking characteristic of the relationship between output—as measured by sectoral gross territorial product—and the number of workers in the sector, is the mammoth difference between this ratio for the petroleum sector and for the rest of the economy. Output per worker in petroleum was almost 11 times that for the economy in 1950, 13 times in 1958, and between 1958 and 1969 output per worker went up 2.2 times, and for the rest of the economy only 17 percent. Ouput per worker in 1969 in petroleum was 25 times that of the national average. The difference is attributable to the substitution of capital for labor, improved labor-saving technology, and the failure of the sector to expand after 1958.[6]

Next to petroleum was mining. Its ouput per worker increased 15 times between 1950 and 1969. Third was energy and fourth was services. These four sectors all had output per worker ratios above the average of the economy. Starting at the bottom, agriculture registered the poorest performance with an ouput per worker in 1969 of 28 percent of the national average. Transportation and communications, construction, manufacturing and commerce had below average output per worker.

Little meaning attaches to the average when the highest output per worker is 90 times the lowest, but it is noteworthy that manufacturing, usually regarded as making economical use of labor, has an output per worker only 78 percent of the average, below commerce and services. Even output per worker in construction exceeds that of manufacturing for most years. It suggests an inefficient manufacturing industry.

Another noteworthy trend is the stationary level of agricultural employment combined with a tripling of output per worker. This is an impressive record, better in fact than all but the petroleum and mining sectors. It suggests that a high degree of hidden unemployment prevailed earlier and that the urbanization process is drawing out many low productivity agricultural workers. The trend was not even over the two decades. The 1950s registered slow increases but between 1959 and 1969, when agriculture received a heavy dose of political rhetoric and mild doses of protection and capital, output per worker almost doubled.

Output per worker in manufacturing went up 1.8 times in the two decades, but nearly the entire gain was between 1950 and 1959. Between 1960 and 1969 employment increased 96 percent but output per worker declined each year, implying that labor productivity may have declined. A trend similar to that in manfacturing occurred in other sectors as well. In construction, ouput per worker increased to 1963 but then declined. In transportation and communications, output per worker went up and down several times but declined overall by more than one-third. In commerce, output per worker ascended to 1957 and then fell.

The only sectors to display continuous increases in output per worker were agriculture, petroleum, mining, energy and services. Output per worker in this last sector increased up to 1964 and then held steady. Indeed, the declining sectors had sufficient weight that from 1963 on, output per worker for the economy stagnated, contrasting with a doubling of output per worker between 1950 and 1963.

Not too much should be read into output per worker data, although the wide disparities and the downward trend for industry and commerce indicate that Venezuela has had a serious problem in making effective use of its labor force, especially in the relatively slow growth of the 1960s. Lower labor productivity on new investment may have slowed growth. Construction, transportation and communications, and commerce may have acted as a sponge, absorbing more workers as output expanded modestly. The government kept up pressure to maintain high employment, even with unused capacity, and labor unions pushed hard to keep their workers on the payroll.

The culprit was the failure of output to expand rapidly in the 1960s. Firms hired workers but underutilized them because output failed to expand sufficiently to make effective use of them. But government policy, the Labor Ministry regulations, and the unions made it difficult and costly to fire workers. If unemployment is the alternative to low output per worker, then Venezuela, with no welfare program, chose correctly.

The data suggest that the efficiency and skills of the labor force did not improve greatly in the 1960s, but could have if growth had been higher. On the supply side, such problems as lack of capital, lack of effective management, and insufficient motivation and discipline of workers remained. But, even more important, on the demand side investment was low, per capita income was creeping up, and government revenues were not increasing rapidly. Data for the 1970s are not available but they will probably show a resumption of productivity increase as growth picks up again.

Capital per Worker

Each worker had at his disposal in 1950, Bs16,500 in capital, with great variation among sectors. Each worker had Bs25,700 in capital, and an increase of 56 percent, by 1969. From 1950 to 1959 capital per worker increased steadily, reflecting high net investment in the period, and by 1959 it was 64 percent greater than in 1950. From 1959 to 1969 the amount of capital per worker declined 5 percent, a result of the relatively slow rate of capital formation in that decade.[7]

Petroleum led all sectors in the amount of capital per worker, 3.8 times the average in 1950, and 9.4 times the average in 1969. Capital per worker tripled between 1950 and 1961, after which increases were modest. The capital intensive nature of the industry requires large amounts of capital, but during the 1960s disinvestment took place which, along with the declining labor force, accounts for the small changes in the decade.

Agricultural capital per worker increased substantially and was 2.2 times in 1969 what it was in 1950. In 1969 agriculture had more capital per worker than did construction, manufacturing, and transportation and communication. The high priority given to agriculture by the government after 1958, stagnation of the agricultural labor force, counting livestock as a part of capital, and the poor showing of manufacturing explain the greater amount of capital per worker in agriculture than in manufacturing.

The amount of capital per worker in manufacturing increased 95 percent from 1950 to 1961. From 1961 on, the lower rates of net investment, combined with rapid increases in the manufacturing labor force, produced declines in capital per worker. By the end of the 1960s, manufacturing had only slightly above the average capital per worker. Construction had the least amount of capital per worker, reflecting its labor intensive nature. Capital per worker in electric power increased significantly, a result of hydroelectric power facilities.

Transportation and communications, like electric power, is capital intensive. It had 2.9 times the amount of capital per worker as the economy as a whole in 1950, but the ratio declined to 2.3 times the average in 1969. In the 1950s commercial capital per worker tended to increase, hitting a peak in 1961, followed by a steady decline to less than one-half the capital per worker it had earlier. For services, capital per worker increased to 1959 and has declined since. In the three tertiary sub-sectors the increase to around 1960 preceded declines, reflecting the high investment and low labor absorption of the 1950s and the low investment and higher labor absorption of the 1960s. Capital substituted for labor in the earlier decade and labor substituted for capital in the later decade.

Capital subsituted for labor only to a moderate degree in recent years with the notable exceptions of petroleum, mining and electric power. The decline in investment and rapid rate of growth of the labor force in the 1960s prevented the amount of capital per worker from rising. The heavy investment in housing also lowered the amounts of investment available to other sectors. Housing capital stock more than quadrupled between 1950 and 1969, a much more rapid increase than capital stock in directly productive activities.

Unemployment

The failure to achieve full employment and to make the most effective use of its human resources was one of the most serious economic problems which Venezuela has faced since 1950. The cyclical crisis of 1959–1962 heightened the problem. There were three years of unemployment of more than 14 percent, and in ten of the twenty years from 1950 to 1969 unemployment exceeded 8 percent of the labor force. In 1970 the unemployment rate began to go down and through the mid-1970s was less than 6 percent. But, in additon to open unemployment, hidden unemployment also plagues Venezuela. Many workers, nominally employed, did not work much of the time, had extremely low productivity, but were capable of contributing more to output.[8]

In 1950, 107,00 workers had no jobs, and at no time since has unemployment been smaller. Between 1950 and 1957 unemployment averaged 123,000 but tended to increase in the last half of this period. In the six years following 1957 unemployment continued to increase, ascending to 361,00 in 1963. The increments to unemployment increased from 1953 through 1960. The average number of people unemployed during the 1959–1963 recession was 322,000.

In 1961 the increases in unemployment began to decline, although the level remained very high. The peak unemployment came in 1963; the peak rate of unemployment in 1962. Unemployment then declined from 1963 to 1968, went up again in 1969–1970, and declined in 1971–1972. Unemployment in mid-1972 was 180,000, an employment rate of 5.4 percent, the lowest rate in the last quarter century. Since 1972 the rate of unemployment has been moderate, in the range of 6 percent.

For the 1950–1969 period the unemployment rate was 10 percent, representing a large-scale waste of resources. Even if one assumes that an economy such as that of Venezuela has a "normal" unemployment rate of about 6 percent, the loss in these two decades was more than 1.6 million man years of work. The prosperity of the 1970s, however, is reducing this waste to some degree.

All occupational categories felt the effect of unemployment. In 1961, a year of economic crisis, unemployment hit hardest: factory workers and artisans, drivers, salesmen and office employees, and miners. The occupations suffering the least were: farmers, service workers, professionals, technicians and administrators. On the other hand, in 1967, a year of relative prosperity but still with substantial unemployment, factory workers and artisans, as well as office employees and farmers, had high unemployment rates. Drivers and salesmen, however, were back to work. In 1971 manufacturing, construction, commerce and services continued to have substantial unemployment.

In 1961 the rate of unemployment in construction was nearly triple the rate of the economy as a whole. The public works program of the dictator had been cancelled and programs of the new government had not started. Unemployment rates were also higher in transportation and communications, and petroleum and mining, as well as manufacturing. On the other hand, agriculture had the lowest open unemployment rate, and energy and services also had low rates. In 1967 construction unemployment was down considerably but the rate was still high. Petroleum and mining had relatively high rates but the manufacturing unemployment rate was low.

Venezuela must also contend with hidden or disguised unemployment. Many agricultural workers actually work only two or three months of the year, or spend only a few hours a day at their tasks. Many city workers sell pencils and a thousand other things on the streets. Government workers pass the day in an office without actually doing any work. No estimate of this kind of unemployment is possible, but it is widespread in Venezuela. The 1974 measures to man all washrooms and self-service elevators adds to the army of non-working employed.

The unemployment data also do not report a large number of people who are not employed, and whose lack of skills or whose health or condition of life prevent their entry into the labor force: the unemployables. The economy could absorb some of these at extremely high rates of growth. Training programs could assist some of them. The number of unemployables is not known. In 1965 President Leoni estimated the figure at 200,000, about as many as were officially unemployed in the last half of the 1960s. This is probably a low estimate and the actual number may reach 400,000 to 500,000.

The reason for the large bulge in unemployment in 1958–1963 was, of course, the recession. Economic as well as political factors conspired to produce an unfavorable juncture in the economy. The petroleum market became staturated, a dictatorial regime fell; revenue did not increase and both public and private programs were cut back. Great uncertainty and a feeling of insecurity prevailed. Construction was the key in employment. Employment in construction was 140,000 in 1959 and fell to 113,000 in 1961, a loss of 27,000 jobs, nearly one-third of the increase in unemployment in these years. In the longer 1958–1963 period, construction lost 43,000 jobs while unemployment as a whole increased 166,000. On the leading edge in prosperity, construction was also the leading edge of the economy in decline.

More fundamental reasons are necessary to explain the chronically high rate of unemployment (6 to 8 percent) even in the absence of the recession. One reason is the high rate of population growth. The potential addition to the labor force was 55,000 to 70,000 workers in the 1950s, leading up to more than 100,000 in the early 1970s. The actual addition to the labor force in the 1950s was an average of 57,000 a year and in the 1960s, 81,000 a year, both far below the increases required to establish and maintain full employment.

Population movement adds to the unemployment problem. Migration crowds the unemployment into the cities and leads to higher open unemployment rates. For example, in 1971, more than two-fifths of unemployment was in the Federal District and Maracaibo. Those leaving the countryside had employment, even if only at very low productivity tasks. Their appearance in such large numbers in the cities did not lead to less output in the agricultural sector, but it did heighten the unemployment in urban areas. In the trade-off between hidden unemployment in agriculture or open unemployment in the cities, Venezuela has in fact opted for large urban unemployment.

Venezuela has tended to employ the technology and modes of production in use in developed countries. This results in limited employment opportunities in the modern sectors, especially of unskilled tasks. The presence of petroleum permits the importation of capital goods and modern technology. The modern sectors are therefore capital intensive and employ few workers. The most modern equipment and a few skilled workers build and maintain roads. In manufacturing, enterprises use the latest techniques, employing capital equipment for tasks that workers could perform. The extreme case is petroleum and mining where the labor force is tiny, but it prevails also in manufacturing, parts of construction, energy, and to a lesser extent in the tertiary sectors. The modern sectors have a low propensity for absorbing workers, especially unskilled and semi-skilled workers.

The high price of labor also discourages the modern sector from employing more people. The petroleum sector with its high wages sets the tone for manufacturing and other modern sectors. Employers thus have an economic motive to substitute capital for labor whenever possible and the country has the foreign exchange to permit them to continue. The presence of strong unions, bargaining for higher wages and fringe benefits under the threat of strike, and with government support, are still further motives for employers in the modern sector to seek refuge in capital equipment. Strikes for political reasons also threaten employers. In additon, the behavior of the government itself in its development projects and in its current operation, is similar to that of private enterprise.[9]

The modern sectors could absorb more of the unemployed, increase productivity and expand output more, except for the low level of skills of many Venezuelan workers. Indeed, for skilled occupations, the unemployment rate is quite low. The lack of skills hampers industry, delays investment and construction, and discourages potential employers. The high-cost output of industry and its inability to compete effectively, as well as its modest progress in the 1960s when petroleum lagged, is evidence of the failure of the labor force to provide workers with sufficient skill or high productivity.

Only recently has the government tailored its educational and training programs to the needs of a modern industrial society. The curriculum in education remains backward and not until the late 1960s and 1970s did the educational system begin to touch the bulk of the population. Technical training programs have been small and limited in scope. In the 1970s the government is beginning to improve education and to expand technical training. In 1974 the government announced a great scholarship plan under which it will send 10,000 Venezuelans aboard each year to study, particularly in technical matters. Most applaud the intent of the program but many criticize it, arguing that the government should use these resources to upgrade the national educational and training programs.

The officially unemployed, the hidden unemployed, and the unemployables—that is, the unutilized human resources—constitute one of the most serious economic and social problems of the country. It represents a significant loss of production, slows the rate of growth, and is a potential source of economic and political instability and uncertainty. Despite the better priority assigned to achieving a higher utilization of labor and improving its skills and productivity, the problem continues.

Labor Organization

Labor unions have played more of a political than an economic role in their brief history. Under the dictatorships of General Cipriano Castro and General Juan Vicente Gómez no genuine labor organizations existed. Attempts to organize workers were ruthlessly suppressed; these governments regarded unions as communistic and subversive. Union history began in 1936 but even then with a heavy government hand. Then, from 1948 to 1958, General Marcos Pérez Jiménez once again extirpated union influence.

The Labor Code of 1936, as amended in 1945, 1947 and 1966, constitutes the basic legislation, government union activities and treatment of labor. The original legislation permitted union organization and required union recognition. The government established the Ministry of Labor in 1937 and brought the agricultural sector under labor laws in 1945. The governments of General López Contreras and General Medina Angarita permitted increasing unionization and sanctioned the foundation of the present labor movement.

The first union formed was in the petroleum industry. Unions spread quickly through manufacturing and most of the modern economy. In mid-1937 the first national labor meeting took place but no permanent central labor organization resulted. The government became uneasy concerning the growing strength of unions, mostly dominated by communists, and banned a planned 1938 labor conference. The Medina government was more favorable to unions. The number of unions increased and collective bargaining began. The second Worker's Congress, held in 1944, also did not produce a permanent national organization, but Acción Democrática did capture control of the labor movement.

The military-civilian uprising which deposed Medina and brought Acción Democrática to power in 1945 ushered in a period of unprecedented union growth. The number of unions quadrupled, collective bargaining procedures became more effective and affected more workers. Local unions federated into large units along regional and occupational lines. The 1947 third Worker's Congress brought into existence the first central organization, the Venezuelan Worker's Confederation, in which Acción Democrática dominated but with strong communist influence.

The military overthrow of 1948 effectively killed the union movement for the next decade. The Confederation and many other union organizations dissolved and the regime attempted to develop its own system of labor organization. The real union movement went underground and many of its leaders escaped into exile. Even so, the union movement lived clandestinely. In the events of 1957 leading up to the hasty departure of the dictator, labor leaders played a key role and popular labor uprisings were prepotent in the overthrow and in the formation and maintenance of the interim government of 1958.

After the ouster of the dictator in 1958, the 600 unions which had existed in 1948 sprang back to life, and 700 more cropped up instantly. The government reinstituted collective bargaining in 1958 and changed the method, recognizing regional and industry-wide contracts, rather than local contracts. Acción Democrática, which in reality controlled the labor movement, accepted a minority role in the reconstituted Confederation in 1959 in order to establish labor unity. The outbreak of strikes in 1959 and 1960 were not economic measures but rather a manifestation of labor power and unity in the face of the threat of a new military dictatorship.

Labor unity, however, collapsed in 1960 and a coalition of Acción Democrática and COPEI established hegemony in the Confederation and the labor movement. The communists split off or were removed from the Confederation and in 1963 formed their own labor organization, claiming 400 of the 3,400 unions which existed at that time. Its principal strength was among white collar workers and some petroleum and mining workers. Mainstream political parties, however, dominated the majority of the unions.

The labor movement was one of the principal strengths in the regimes of Presidents Betancourt and Leoni, and was instrumental in the election of President Carlos Andrés Pérez in 1973. Collective bargaining spread in the 1960s and 1970s and became the accepted means for wage determination for most of the economy. The government and the people accepted unionism fully although many complained of its inordinate economic and political power. The Confederation is the most powerful national labor organization, with the smaller communist and Catholic organizations playing a minor role.[10]

The union framework in the 1970s is formally similar to that which exists in developed countries. The government denies the right to organize only to its own white-collar workers and the military. Anyone else may join a union, but membership cannot be a condition of employment. Bargaining contracts made between union and employer are applicable to all workers. No one can be refused employment because he is a union member.

About 2,000 local unions existed in 1964 and since that time new unions have been formed at the rate of about 300 per year. Most of the local unions are quite small and nearly all are members of a federation. In

1969, for example, the government reported 342 new unions with a membership of 17,000. In that year 143,000 new people entered the labor force. The Confederation claimed a membership of over one million in 1970, about one-third of the labor force, organized into 23 regional and 24 industrial and trade federations.

As economic organizations, the industrial and trade unions are more effective than other parts of the labor movement. About 45 percent of the membership of the Confederation is in one union, the Agricultural Worker's Federation, which had 3,500 locals and claimed 600,000 members. The total labor movement claims 1,300,000 members, which, except for the agricultural bloc, tend to be in the industrial sectors in the cities. One of the strongest unions has been the 20,000 member Worker's Petroleum Federation.

The law requires that labor unions and political parties be independent of one another, but close cooperation and coordination has always existed between many unions and the parties to which most of its members adhere. Labor leaders are usually also important figures in political parties. The interaction between unions and political parties has tended to strengthen both, especially the party in power, since that party necessarily has a strong labor element to get elected. This has been true of the governments of Acción Democrática and COPEI as well.

Unlike the situation in many developed countries where the principal goal of unionism is the maximization of the economic benefits to its members through the collective bargaining process, in Venezuela an equally important goal is to achieve economic benefits through political influence in partisan politics. Nearly all unions are highly politicized and divisions within the unions most often occur along party lines. Political motivations account for much of the labor unrest and strikes. There are relatively few strikes and when they do occur they are seldom for the purpose of bettering the collective bargaining agreement.

Personalism, personal popularity and influence, and alliances based on friendship and kinship characterize union organization. The performance of the union and union leadership is not based upon advantageous collective gargaining agreements or on securing additonal pay or benefits for union members, but rather on the connections the unions and their leaders have in the government and in the labor movement itself. The founding generation of the labor movement, however, is gradually relinquishing its leadership and unions are beginning to behave more like economic organizations.[11]

Education

Education in Venezuela has not been well suited to maximizing the contribution of the country's labor force to economic development. It has, however, improved, and has served and is serving several critical functions. The purpose of the vastly increased educational effort of the 1960s and 1970s has been to generate new skills and to increase productivity. It has also served to withhold people from the labor force, retaining them in educational system rather than permitting them to join the ranks of the unemployed. Upgrading the productivity of the existing labor force through the diffusion of literacy and through adult educational and training programs has been a third role of greater education. A fourth function is that of improving the general level of economic, social, political and cultural sophistication.

Since 1958 the contribution of education to changing the system has been positive. Before 1958 education was a difficult stepchild of the regime. The national educational budget has since increased more than 20 times. The share of the budget going to education in 1950 was 6 percent; in 1974 it was 19 percent, higher even than in most developed countries. In 1974 nearly 27 percent of the population was in the educational establishment, whereas the rolls in 1961 included only 17 percent. And this does not count an incredible amount of adult and specialized education. The spectacular growth, however, has most often been quantitative.

Improvements in the quality of education and in directing education to the labor market have been less notable. The Ministry of Education undertook, in the early 1970s educational planning and curriculum revision, and the introduction of changes has begun. The reforms, however, will require many years to effect, and their impact to be assessed. Even the numerical growth of the system has exceeded facilities. In 1969, for example, the government constructed new school buildings for 42,000 more students, but new entries were 118,000.[12]

The government has assigned great emphasis and high priority to all levels of education but the focus has been at the secondary level. The number of secondary school teachers tripled between 1958 and 1969 and the number of students grew even more rapidly. The component which emphasized technical training experienced the greatest growth of all. In additon, the National Institute of Educational Cooperation, formed in 1958 as an organization outside the school system, has the mission of providing technical, vocational, and on-the-job training. Its growth was rapid in the 1960s. Moreover, technical training in commerce, services and agriculture is available to an increasing degree. Universities have expanded, branches and specialized facilities have been established, and new universities have sprung up. Universities have begun to change their orientation, away from law and the humanities toward technical and scientific fields, business administration, forestry, agronomy and specialized fields of petroleum.

Two problems continue to plague the educational effort. The dropout rate remains very high. This is particularly true after the third and sixth years and every year of university education. The dropout rate in rural areas is twice as high as in urban communities. The other problem is the still outdated curriculum, a carry-over from the Franco-Spanish tradition, with much emphasis on rote learning and little focus on analysis. Curriculum reform is in process, and different educational levels are being integrated, but the emphasis on quantity has outstripped improvements which would have rendered the entry of students into the labor force even more productive.

Adult education made substantial progress in eliminating illiteracy and permitting many to complete their formal education. The rate of ilteracy was 53 percent in 1950; by 1968 it was 81 percent of the population over 10 years of age. Special literacy teams worked in a concentrated effort to teach people the basics of reading and writing. It was successful among males in urban areas. In addition, adults had available all levels of education through regular and accelerated programs in regular school facilities at night, not only in academic fields, but also in technical and vocational subjects. Its purpose is to permit workers to complete the schooling they missed earlier and to make them more productive. More than 100,000 adults were in this type of special education at the beginning of 1970.

While clearly more people receive more education and training of all types, the productivity of the labor force was yet to demonstrate marked improvement. Like other programs of economic development, an educational effort, even a large-scale one as in Venezuela, requires a considerable period of time to bear fruit. The relatively short life of the accelerated program, the lag in curricular and institutional reform, and the failure to direct education to the specific manpower needs of the economy has so far blunted its effect.

NOTES

1. The data on population and its economic and social characteristics comes from the Dirección General de Estadística y Censos Nacionales and is reported in census reports, the tenth of which was taken in November 1971, and in the *Anuario Estadístico*. The reliability of early census data is open to question, but in recent

years they are regarded as reliable; see Alfredo Rodríguez. *La Estadística en la Historia de Venezuela*, Caracas: Ministerio de Fomento, 1973.

2. The volume by José Eliseo López. *Tendencias Recientes de la Población Venezolana*, 1968, is a useful study of population trends and characteristics.

3. Chi-Yi Chen studies migration in his *Movimientos Migratorios en Venezuela*, 1968; for the relationship of population to development in Venezuela, see Tómas Enrique Carrillo Batalla. *Población y Desarrollo Económico*, Caracas: Banco Central de Venezuela, 1967.

4. *Anuario Estadístico 1972*, pp. 55–86; and *Venezuela en 1972*, Cuarto Mensaje del Presidente de la República, Caracas: Oficina Central de Información, 1973, Tables 4–14.

5. The *Anuario Estadístico* contains detailed data on the characteristics of the labor force.

6. The output figures employed in this section are taken from the national accounting data of the Banco Central de Venezuela as published in *La Economía Venezolana en los Ultimos Treinta Años*, 1971, in 1957 prices. The change in the price base for data after 1969 limits comparisons.

7. Data on capital stock are from the annual *Informe Económico*, Banco Central de Venezuela and *Venezuela, 1950–1967*, 2nd edition, Washington: Inter-American Development Bank, 1968. Capital stock figures are in 1957 prices.

8. Unemployment data are in *La Economía Venezolana en los Ultimos Treinta Años*, 1971; and the *Anuario Estadístico*.

9. Mostafa Hassan. *Crecimiento Económico y Problemas de Empleo en Venezuela*, Caracas: Banco Central de Venezuela, 1973, analyzes employment and growth, especially in chapters IV and V.

10. See Chi-Yi Chen. *Economía Social de Trabajo: Caso de Venezuela*, 1969; and Robert Alexander. *The Venezuelan Democratic Revolution*, 1964.

11. Daniel H. Levine. *Conflict and Political Change in Venezuela*, 1973; John Martz. *Acción Democrática*, 1966; and John Duncan Powell. *Political Mobilization of the Venezuelan Peasant*, 1971, analyze the labor movement and politics.

12. The Dirección General de Estadística y Censos Nacionales compile educational data and report them in the *Anuario Estadístico*. For a discussion of education in Venezuela, see Carlos Acedo Mendoza. *Venezuela: Ruta y Destino*, 1966, chapter X; and George I. Sánchez. *The Development of Education in Venezuela*, Washington: U.S. Department of Health, Education, and Welfare, 1963.

Chapter IX

Money, Banking and Finance

The financial system of the Venezuelan economy has grown rapidly, matured in its institutional framework, and gained flexibility and sophistication in the last four decades. In 1936 banking was primitive, served few clients, and restrained development. Banking and finance were still in a rudimentary stage in 1950, although greatly expanded quantitatively. Bill and coin circulation exceeded demand deposits. The Central Bank of Venezuela was only ten years old and had limited legal instruments in pursuing monetary policy. The institutional structure was skimpy, and aside from commercial banks, few financial institutions existed. The few commercial banks provided elementary services to industry, commerce and the economic oligarchy.

In the mid-1970s Venezuela had one of the most advanced and complex banking and financial communities in Latin America. By the end of 1975 the value of time deposits were 20 times their value in 1950, and more than 6 times their 1960 value. Time deposits were up over 80 times 1950 deposits, and nearly 8 times 1960 deposits. The number of banks had doubled and through their branches enveloped the country in an integrated monetary system. Specialized financial institutions such as savings funds, investment funds, savings and loan associations, mortgage banks, public financial institutions—headed by a strong Central Bank—complete the system of financial intermediaries.

Money Supply and Liquidity

Venezuelan monetary authorities define the money supply as coins, bills (issued only by the Central Bank) and demand deposits. Monetary liquidity is: money supply plus time deposits. The change in monetary liquidity in any given period of time is the sum of the credit, fiscal and exchange effects. The credit effect is the net amount of private credit added by net lending by the banking system, independent of government and international transactions.[1]

The fiscal effect registers the difference between government revenues and expenditures which flows through the banking system; it is public credit. If expenditures exceed revenues, the government, in effect, is putting money into circulation, either out of reserves or by borrowing. If revenue exceeds expenditures, the government is reducing liquidity by withdrawing funds from the system. The exchange effect is the amount which clears the international accounts, the balance of payments; it is international credit. A positive exchange effect means that more foreign exchange is entering the country than leaving on the consolidated international accounts. The algebraic sum of the three effects is the change in monetary liquidity.

The underlying causes for the expansion or contraction of money and liquidity are changes in the supply of foreign exchange, the fiscal position of the government and the net bank credit extended. These are interrelated through a complex set of financial flows. An increase in foreign exchange adds to liquidity, as does a government deficit, as well as an increase in private credit. A decrease in foreign exchange, a government surplus, and net liquidation of bank credit results in a loss of liquidity.

In 1950 the money supply consisted of coins, 6 percent; bills, 44 percent; and demand deposits, 50 percent. The coin participation has tended to slip throughout the period, as have bills.[2] Bills were 27 percent of the money supply in 1972. Demand deposits grew rapidly and in 1972 were 70 percent of the money supply. In October 1975, they reached Bs12,500 million, compared to Bs684 million in 1950, Bs2,167 million in 1960, and Bs4,735 million in 1969.

The growth of demand deposits has been volatile. They increased from 1950 to 1958, with especially large increases in 1956–1958. Deposits then fell in 1959 and 1960, increased slightly in 1961, fell again in 1962, and then increased in 1963, reflecting the vicissitudes of the recession. It was 1964 before the deposits exceeded the 1958 level. A slight decline occurred in 1966, but growth has been sustained since, with large increases each year. In 1973 growth was 29 percent, and in 1974 it was 44 percent.

Total money supply in 1950 was Bs1,558 million, and in 1974 it was Bs14,969 million. The increase was steady from 1950 to 1958 when money supply was Bs3,856 million. A decline followed until 1962, when it was Bs3,382 million. The recovery began in 1963; in 1964 money supply exceeded the pre-recession level and increases have been continuous since, nearly doubling by 1970 and more than doubling between 1970 and 1974. The importance of the trend of government receipts from petroleum is clear in the expansion of money supply.

Time and savings deposits, a part of monetary liquidity but not money supply, had an equally—and in some years a more—impressive record than demand deposits. These deposits increased through 1959, declined in 1960 (a year later than in the case of demand deposits) and began to recover in 1963, a year earlier than demand deposits. With only a hesitation in 1966, they have grown spectacularly since. In the mid-1960s time deposits caught up with demand deposits and have grown more rapidly. In 1974 time deposits were 12 percent greater than demand deposits, even though in that year demand deposits went up 44 percent and time deposits only 22 percent, reflecting the petroleum windfall.

Monetary liquidity changes consolidate all of these trends. From 1950 to 1958 monetary liquidity increased steadily, but in 1959 remained almost constant. In 1960 it took a sharp plunge, going down by 10 percent. In the following three years there were small increases, but in 1963 the increment was large, large enough so that monetary liquidity exceeded the 1959 level. Except for the standstill year of 1966, subsequent additions to monetary liquidity have been substantial since 1963, with an increase of 20 percent in 1971, and 18 percent in 1972. In 1974 monetary liquidity was 64 percent greater than in 1972.

Each of the sources of change in monetary liquidity has shown great movement in recent years. Private credit change has always been positive, but during the recesion it was almost zero. The fiscal effect has tended to be negative, with a surplus of revenues over expenditures, thus reducing liquidity. In the recession years deficits, however, occurred, as they have in some other years. The exchange effect has moved from strongly positive to strongly negative, adding liquidity in 1956–1957, in 1971 and subsequently, and in a few other years. During the recession, however, net capital outflows reduced liquidity.

The trend in liquidity was upward from 1950 to 1955, largely because of private credit expansion. In 1956 and 1957 there were strong government supluses, more than offset by capital inflows and a favorable private credit market. In 1958–1960 the private credit market all but disappeared, the government ran deficits, and capital flowed out. Additions to liquidity declined and liquidity itself went down in 1959 and 1960. In

1961 the capital outflow ceased, the government ran a deficit (small compared to those earlier) and the private credit market hit rock bottom; liquidity stood still. In 1962 a large government surplus offset a substantial expansion of private credit so that liquidity was again nearly constant.

The recession overcome, liquidity expansion resumed in 1963, largely because of private credit, offset in part by government surpluses and an occasional capital outflow. In 1966 the capital outflow plus the government surplus offset private credit and liquidity was almost constant. Since 1966 the expansion of liquidity has been impressive, with rapid expansion of credit, and mixed fiscal and exchange effects, both primarily adding to liquidity. Since 1970 the exchange transactions have added greatly to liquidity.

Much of the behavior of money supply and liquidity is attributable to factors outside the control of the monetary authorities and the banking system. Private banking has been a significant element in the expansion and it is to some extent responsive to policies of monetary authorities. Even private credit, however, is sensitive to petroleum exports as a barometer of economic prospects and to the state of the government budget. The latter, of course, is directly dependent on petroleum revenues and hence exports. Petroleum exports in turn depend on the demand of the industrialized countries and on government policy. For these reasons, the Central Bank has only a limited influence over money supply and liquidity.[3]

When petroleum revenues began to flow into the country as a result of price increases in late 1973, Venezuela confronted a very rapid expansion of liquidity through international transactions at a time when private credit expansion was strong. The government wished to sop up some of this liquidity to prevent too great an expansion. It established the Venezuelan Investment Fund in mid-1974 to sequester some of the revenues. In the year following its foundation in July 1974, the government set aside Bs19,621 million, nearly one-half of the incoming revenues, placing the funds in international capital markets.[4]

Structure of the Financial Community

Private banks dominate the financial scene in Venezuela. At the end of 1972 there were 31 commercial banks with 198 branches and 362 agents scattered throughout the country. More than 500 other financial institutions exist, but more than 400 of these are small savings funds. In the early 1970s commercial banks were responsible for nearly one-third of all financial flows, and private financial institutions for about one-half.

The public sector handles the other half of financial flows. The largest institution is the Central Bank which serves as the government's bank, the private banker's bank, and the bank for financial institutions in the public sector. Three other government financial institutions are important. The Venezuelan Development Corporation promotes private development through loans and stock purchases. The Agricultural and Livestock Bank lends to agriculturalists. The Worker's Bank is a low-cost housing lending institution. The Venzuelan Investment Fund is a newcomer which now has all its assets in international capital markets, but over a period of time it will finance development projects within the country.

The trend during the 1960s and 1970s was toward greater public sector direct participation in the financial system. Except for its handling of foreign exchange, which increased dramatically in the 1970s, the Central Bank participation tended to decline. The greater public sector participation in finance arose because of the creation of new and specialized institutions, funded partly or wholly by the government, as well as increasing financial activities of existing institutions. The Guayana Corporation of Venezuela, the Worker's Bank, and the National Savings and Loan Association, and other governmental and semi-governmental bodies are increasingly important sources of finance.

The institutional arrangements in the private sector experienced a major change. In 1950 there were only 16 commercial banks with 18 branches and 31 agents. By 1960 there were 37 banks, 150 branches and 286 agents. In the 1960s the growth of banks halted and some consolidation occurred so that in the 1970s there were fewer banks with a larger number of branches and agents. But in the 1960s and 1970s new kinds of institutions came into being, including mortgage banks, insurance companies, savings and loan associations, investment funds, and savings funds.

Commercial banks serve primarily short-term credit needs of commerce and industry. Most loans are for 30, 60 and 90 days, nearly all less than two years. Recent liberalization has encouraged commercial banks to lend for longer terms and to agriculture. Mortgage banks, savings and loan associations, and government banks, such as the Worker's Bank, make long-term housing loans and the new investment funds lend for commercial and industrial expansion. Even the commercial banks provide longer term loans by the expedient of rolling over short-term loans.

Central Bank of Venezuela

The Central Bank of Venezuela, authorized in 1939, began operations in 1940. Before that time commercial banks, especially the Bank of Venezuela, acted as fiscal agents for the government and performed limited monetary functions. Only 11 commercial banks existed, of which the Bank of Venezuela was the largest. Each bank had its own notes and a small clientele. Despite the minimal monetary activity, the currency was strong, there was no debt, and banks followed a conservative lending policy.[5]

The establishment of the Central Bank centralized the emission of bills and the Bank accepted the responsibility for managing the foreign exchange system and international reserves. It became the central depository of the reserves of the commercial banks, which had to keep at least two-thirds of their reserves with the Bank.

The Bank, however, did not have all of the attributes of a modern central bank. The reserve ratio was in the hands of the Congress and the Bank could engage in open market operations only under such restrictions, that in fact the Bank did not operate in the market. It could rediscount commercial bank paper, but this function also had limitations. At its initiation, the Bank was more the shell than the reality of a central bank.

The government created the Central Bank as a private company with ownership divided equally between the government and the private sector. Limitations prevented an individual from owning more than a small fraction. In 1975 the government took over full ownership and complete control. A National Banking Council advises the Bank. The Bank president and senior officials work closely with the ministries, especially the Treasury Ministry. The Superintendent of Banks, an office in the Treasury, has charge of the legal and technical supervision of banks and of enforcing the law. It inspects, audits and controls all credit institutions but has no monetary policy functions.

The 1940s were learning years and no grave crisis disturbed the steady expansion of Central Bank assets. The presence of the Bank was itself a radical departure from the past to render a change in monetary thinking and practice, even without vigorous activity by the Bank. The Bank interpreted its role as the conservative management of government money, to maintain vigilance of the monetary and credit system, to keep the bolívar sound and stable, and to establish its own credibility. Officially the Bank had a development role but its direct participation in finance was limited. Even acting in this passive role, the Bank supervised a great expansion of money supply and monetary institutions.

In 1950 the assets of the Central Bank, more than 10 times that on hand at the Bank's founding, consisted almost exclusively of the country's international reserves. Throughout the 1950s Bank assets continued to be 85 percent or more in the form of international reserves. Loans and investments of the Bank were 10 percent or less of assets. Assets expanded rapidly from 1950 to 1957, then declined precipitously from 1958 through 1960.

The composition of Bank liabilities changed significantly in the decade of the 1950s. Bill emission increased substantially, doubling in the decade. On the other hand, deposits were highly variable, largely because of changes in Treasury deposits. Between 1950 and 1955 Treasury deposits doubled, but in the next year they tripled, and in 1957 continued to increase rapidly. The unusual activity in the government account arose because of concession payments. In 1958 and 1959 the government used up these deposits, paying off debt and mounting an emergency anti-recession plan. Deposits by commercial banks were also inconstant, going up the first two years of the decade, then down, then up again from 1954 through 1958, then down again the last two years of the decade.

Clearly the international reserve position of the country was calling the tune for the Central Bank and for finance generally through the 1950s, as modified by the government's fiscal position. The government ran small surpluses during most of the 1950s, except in 1956–1959. The first two of these years witnessed large budget surpluses, offsetting in part the effect of incoming reserves. In 1958, however, while reserves were declining, the government ran substantial deficits. The loan portfolio of the Bank was small and highly variable. The role of the Bank in the 1950s continued passive, adjusting to the international and fiscal position of the country and exercising little independent influence on the financial system.

The boom and bust of petroleum, the end of the dictatorship, and the uncertainty in the changeover to the new regime led to the abnormalities of 1956–1960 and to the recession and monetary contraction of the early 1960s. While it was necessary to adjust to changes in international reserves and the government's fiscal position, the Central Bank did little to restrain the boom. It did, however, try to offset the financial decline and to operate as a counter-weight in the economy. In 1958 private credit expanded by less than one-half that of the previous year. The government, however, ran a deficit, pumping new money into the economy and offsetting the flight of capital.

In 1959 the Central Bank expanded its own credit substantially, began to rediscount loans, and made advances to the commercial banks. Private credit expanded somewhat and the government ran another deficit. The capital flight, however, continued, and the net result was a reduction in Bank assets for the second year, a reduction in the money supply, and no change in liquidity. Then, in 1960, despite another expansion of Bank loans and investments, private credit collapsed, the flight of capital continued, and Bank assets declined again, in spite of the government deficit. The Bank's efforts to pump reserves into the banks was feeble compared to the forces leading to contraction.

It is easy to be critical of the Central Bank for not acting more vigorously to offset the decline which began in 1958, and the government for not enabling and forcing the Bank to participate in a coordinated fiscal and monetary policy which might have stopped the recession. It is appropriate to remember, however, the fiscal orthodoxy of the government at this time, and to consider that many of the influences bearing on money supply were simply beyond the control of both the Bank and the government. In addition, the Bank had limited legal instruments of policy until the change in the Bank law in 1961. Nonetheless, more coordinated Bank-government policies and more aggressive actions on the part of both would have softened the impact of the recession and hastened the recovery.[6]

The Bank made even greater advances to the commercial banks in 1961, and rediscounting went up sharply. Even more important, in that year the government ran only a small deficit and few international reserves were lost. Hence, despite the failure of private credit to increase, the monetary contraction halted. Financial recovery had not begun, however. Central Bank international assets hit bottom, with a further loss of reserves, but other assets, based on advances and rediscounts increased enough so that there was a modest increase in total assets.

The Central Bank position in 1962 was nearly the same as in 1961. There was a small increase in total assets, almost no increase in international assets. Compared, however, to the previous three years, it was a good year because the decline in international or in total assets halted. Overall change in liquidity hovered near zero despite a record extension of private credit, second only to 1957. The government, however, ran a surplus of the same magnitude as the deficit of 1959, reducing the money supply to the pre-1957 level.

The recovery began in earnest in 1963. Central Bank assets have climbed every year since that date, except 1966, when they were stationary. International assets have also increased each year, except 1966, when they declined 9 percent. The Bank has continued to make advances and to rediscount paper to a much greater extent than before 1959, but at lower levels than during the recession. At the beginning of 1972, Central Bank assets were nearly 6 times greater than in 1950. The gains in international reserves in 1973 and 1974 have subsequently greatly expanded Central Bank assets.

The composition of Bank assets has changed. International reserves were 91 percent of total assets in 1950; the ratio increased to 96 percent in 1957. In 1962, on the other hand, the last recession year, international assets were down to 66 percent of the total. By the early 1970s, however, the ratio was back up to more than 90 percent. On the liability side, bills emitted have declined in importance as deposits, not only of commercial banks, but also of specialized institutions and public sector institutions, have increased greatly.

The Central Bank has the usual policy instruments, including rediscounting of commercial bank paper, advances and loans, open market operations, and control of reserve requirements. It can set maximum and minimum interest rates. The banking reform of 1975 added to its policy arsenal. The open nature of the economy, and limited capital market, reduce the effectiveness of monetary policy. Most banks maintain excess reserves and seldom need rediscounts or loans. Even the 1975 increase in reserve requirements does not limit banks. The 1973 reform of the capital market has not yet produced an open market in which the bank can effectively control the money supply.

Commercial Banking

Commercial banking got a late start in Venezuela. Despite various efforts from the beginning of the national period, it was not until 1877 that banks and credit institutions began to appear and the first presently existing bank was not established until 1882. Even then, progress was slow and banking was weak and unimportant. In 1915, for example, bank deposits were only Bs15 million. Even as late as 1940 deposits were only Bs217 million and loans and investments were Bs183 million. The 1940s, however, witnessed a substantial expansion in the number of banks, deposits quadrupled and loans increased 4.6 times in the decade.[7]

The decade of the 1950s was the flowering of commercial banking. Nearly every indicator of bank activity went up continuously, although signs of trouble appeared in 1959 when demand deposits dipped. Bank formation was popular; 21 new commercial banks appeared, as well as 126 branches of old and new banks. Banks established 273 new agencies around the country. By 1959, for the first time, banks operated throughout Venezuela, in every urban center of importance, although Caracas and Maracaibo remained the financial centers of the country.

The record of growth in the decade was impressive. Between 1950 and

1959 loans and investments expanded nearly six times, while all assets expanded more than five times. Demand deposits quadrupled up to 1958. Time and savings deposits were the fastest growing liability, multiplying 13 times in the decade. Even paid-in capital and reserves went up 4.7 times. Each year registered another new all-time high in almost every entry in the books of commercial banking. Banking services, so long lacking or inadequate, proliferated, became more competitive, and banking improved vastly.

The rapid expansion of bank activities during the 1950s, and the increased credit helped to support the high rates of investment during the decade. In the late 1950s the amount of bank loans outstanding was about the same as all non-extractive activity investment. The banks, of course, did not finance all of that investment, but they were operating at a high level and supporting a credit expansion which made the investment possible. The average rate of increase in loans in the last five years of the decade was 23 percent, with one increase, in 1957 of 44 percent. Additions to deposits kept pace with the credit expansion.

Even in 1959 when the signs of trouble began to appear, the rate of loan expansion declined only slightly, national balances declined somewhat, and the ratio of required to actual reserves jumped as the latter fell off significantly. Other accounts on the asset side also began to go down. On the liability side demand deposits declined slightly, although savings and time deposits increased. Overall assets and liabilities in 1959 increased by 5 percent, the slowest increase of any year in the decade.

In 1960 the bank expansion came to an abrupt halt and the contraction set in. In 1960 assets and liabilities declined for the first time. All asset accounts went down, as did total deposits, with demand deposits the hardest hit. It was not a collapse of commercial banking but it was a significant setback, the first since the worldwide depression of the 1930s.

The commercial banking system did not operate independently of the internationally oriented matrix of conditions in which the entire country functioned. Dollars and other foreign currencies could readily be purchased with bolívar deposits. The banking system handled the entry and exit of capital, and with complete convertibility, which prevailed except from 1961 to 1963, no method existed for stopping the movement of capital. When the inflow of capital from the concessions, and the high priced and increased volume of petroleum exports ceased in 1957, the large scale exit of capital followed immediately in 1958 and 1959. The banks could have taken this transfer of money abroad as a signal to restrain lending, since some of the lending, directly or indirectly, was financing the flight of capital. They did not.

In 1958 the increase in liquidity fell sharply and in 1959 liquidity went down for the first time. The banks again, however, did not accept this signal since at the same time that liquidity was going down on the international account, the government was injecting liquidity on the fiscal account. The budget deficit more than offset the international loss and the banking system went on expanding. In 1959 the budget deficit was cut sharply but the banks continued to expand, some of which again went to finance the outflow of capital. Regardless of the actual mechanism of financing the capital outflows of 1958 and 1959, private credit creation by commercial banks acted as an offset to the international flows, which would have been lower had it not been for the new credit.

When private credit did dry up in 1960 the capital outflow declined. The government deficit was not enough to offset the outflow so that a reduction in liquidity occurred. If credit creation in 1960 had continued at the pace of earlier years, the capital outflow would have been greater, since exchange was freely convertible until November 1960. It was the year of monetary crisis, and the entire banking system moved backwards.

In 1961 commercial bank assets and liabilities increased, but the year was not much of an improvement over the previous one. Increases in loans and investments was small, deposits were stagnant and banks received advances and rediscounted at the Central Bank. Even 1962 was not much better. For three years, 1960–1962, the commercial banking system vegetated. While private credit improved in 1962, the government deficit financed it and deposits did not increase.

The openness of the economy made it inevitable that commercial banks expand private credit to pay for the outflow of capital. People wanted to borrow to finance balances abroad. When exchange controls became effective in 1961, private credit collapsed because no one wanted to borrow for domestic purposes and even the 1962 private credit was based on the government deficit. International forces and the government buffeted the commercial banking system for three years. The banking system had no will of its own and monetary policy was no guide.

On the other hand, the commercial banking system stood up under severe pressure. No banks failed. The conservatism of the banks in lending policy and excess reserves saved them from asset losses which might have endangered or toppled a more risk-prone system. While commercial banks made little positive contribution to recovery, neither did they worsen the situation. They did finance the capital outflow, but in fact the government mandated this course by maintaining free convertibility and by paying foreign-held debt.

The recovery of commercial banking was rapid from 1963 to 1965. In those three years total assets, as well as loans and investments, expanded by 35 percent and deposits by 48 percent. The banks were more nearly loaned up in 1965 than at any time in their history. The required-actual reserve ratio was 80 percent. In the first two of the three years capital flowed in, which, added to the credit expansion, more than offset government surpluses. The year 1965 witnessed the largest private credit expansion of any year in the 1960s, but it was offset in part by a capital outflow and government surplus.

In banking, as in other parts of the economy, 1966 was a standstill year. The outflow of capital, plus the government surplus just offset the expansion of credit. Nearly all of the accounts of the commercial banks registered only slightly in excess of the previous year. This was the year when everyone, including the commercial banks, awaited the outcome of the government-petroleum-private sector confrontation on planning, taxes and international economic policy.

In 1967 the expansionist movement resumed and has continued since. From 1966 to 1972 the assets of commercial banks more than doubled. Loans and investments also doubled, and total deposits more than doubled. Even though the late 1960s and early 1970s were not marked by high prosperity, the banking community was expanding. The expansion moved into high gear in 1973 and 1974. Demand deposits increased 77 percent, time and savings deposits 46 percent. Loans and investments were up 69 percent. Banks were liquid and had plenty of excess reserves.

The most important bank is the Bank of Venezuela; it is one of the oldest and most prestigious. Before the formation of the Central Bank it was the government's bank and it still receives the government's revenues. Among the other large banks are the National Discount Bank, the Merchant and Agriculture Bank (formerly affiliated with Chase Manhattan), the Union Bank (founded in 1954 but one of the most rapidly growing banks), the Caracas Bank, and the Commercial Bank of Maracaibo, one of the few large banks not headquartered in Caracas. Two foreign banks, United Holland Bank and National City Bank, continue to function, but not as regular commercial banks.

Banking has evolved from an aristocratic club and the exclusive concern of the rich, into a competitive business enterprise, with each bank eager to capture more demand and savings deposits and ready to make loans on sufficient collateral. The modernization, however, is not complete. Just beneath the surface is the old bank conservatism. If one wants a loan, it is important to know a bank officer or a major client. Old friends receive favored treatment.[8]

Despite the expansion, commercial banks are only now beginning to depart from their traditional moorings in lending policy. Banks have not engaged in long-term lending and until recently they have been prohibited by law from lending for more than two years. They achieve some of the effect of long-term lending by continuous loan renewal, but this is high cost and uncertain for the borrower. Banks have shied away from agriculture and industry, preferring commerce. They operate, for the most part, on the basis of the old commercial banking theory, financing goods in process for short periods. It is profitable, secure and liquid, even though it restricts the development role of banking.

Commercial banks have performed a development role and avoided the long-term lending restriction by equity holdings in enterprise. No limitations exist on bank purchases of stocks in enterprises—and they are a significant proportion of bank assets. During the 1950s commercial banks expanded their connections with nearly every kind of manufacturing, commercial and service activity through stock ownership. The "other accounts" category increased 13 times from 1950 to 1958. This type of operation fitted in well with the Venezuelan style of banking, which emphasized personal contact and involvement in the affairs of clients.

The attitude toward reserves characterizes the conservatism of commercial banks. In 1950 the reserves of commercial banks were Bs275 million, even though only Bs136 were necessary under the law. The banks could have doubled their loans without limitations by the legal reserve requirement. Throughout the last quarter century banks have maintained excess reserves. In 1974 they reached Bs1,117 million, 24 percent more than necessary. In 1975 the Central Bank received authority to change the reserve requirement and has increased it.

The banks also have insisted on heavy collateral, sometimes several times the value of the loan, and have exacted relatively high rates of interest, beyond the maximum eligible for rediscounting. Rediscounting with the Central Bank has been uncommon, for several reasons. The Bank specifies maximum rates of interest on loans eligible for rediscounting. Since many loans earn higher interest, the commercial bank cannot rediscount them. The commercial banks always have excess reserves if they wish to make new loans. Moreover, since commercial bank policy is conservative and profitable, banks have no trouble borrowing abroad in case of need. Finally, the banks themselves tend to interpret rediscounting as a sign of weakness and look upon a bank which rediscounts as a bank in trouble.[9]

The accelerated expansion of the 1970s brought with it changes in the institutional structure of banking. Legislation in 1970 and 1972 altered the role of foreign and foreign-owned banks. While foreign banks date back to 1916, they have not played a major role in commercial banking. The recent laws require that at least 80 percent of a bank must be owned by Venezuelan nationals or else it cannot exercise commercial banking functions. Several Venezuelan banks had greater foreign ownership but nationals have bought out foreign interests and they continue to operate. First National City Bank, however, continues to function as a branch bank and therefore has lost privileges.

Specilized Financial Institutions

While commercial banking dominates the country's financial operations, a large number of other institutions perform specialized financial functions. Ten mortgage banks and 22 savings and loan associations lend money for housing and emit savings instruments which have become popular in recent years. Mortgage banks did not begin until 1958, and savings and loan associations first started in the mid-1960s. Many firms and government agencies have small savings funds. In addition, 39 insurance companies operate in the country and engage in lending to industry, commerce and for housing. Investment funds have grown rapidly in the 1970s.

Aside from the commercial banks, the most important private financial institutions were the mortgage banks and savings and loan associations. Mortgage banks loans increased 3.7 times between 1970 and 1974, when such loans were 30 percent of the loan portfolio of the commercial banks. Savings and loan operations have also expanded rapidly but their scale of activity is only about one-half that of mortgage banks. Mortgage bank money comes from issuance of bonds, while savings and loan money comes from savings deposits.

The many other, smaller financial institutions of the country do not control massive resources but do perform essential specialized functions. The small savings associations, for example, are an important cache of savings which would not otherwise exist. The new investment funds are becoming a significant source of funds for new private ventures. Insurance companies play a vital role in real estate mortgages, public debt, and long-term industrial loans. Short-term financing of consumption has become common in recent years through credit cards; Venezuela has them all, Master Charge, BankAmericard, Diners, American Express. Commercial banks sponsor most cards. The many and varied financial institutions, most of which are new, round out and bring the Venezuelan financial system near to modernity and completeness.

The private financial system of Venezuela in 1950 rested on the slender base of a few commercial banks. During the 1950s the assets of these banks expanded rapidly, but it was not until 1958 and after, that any significant institutional changes emerged. The commercial banking system successfully weathered the recession, partly because of its conservatism and high liquidity, and partly because the recession did not deeply threaten the value of their assets. The recession did demonstrate the narrow financial base of the economy, and during the 1960s, in addition to its expansion, the financial community developed new institutions and intermediaries for channeling resources into industry, commerce and real estate, and for the mobilization of savings.

Despite the progress in private finance some of the problems of tradition and inflexibility still plague its operations. Commercial banks remain conservative, oriented to short-term loans to clients known personally to the bank. The newer institutions also tend to operate on a friendship and kinship basis. The capital market is still gravely deficient. The long-term credit needs of the economy are still not met through private finance. Finally, the financial community remains highly vulnerable, in the regime of free convertibility, to international influences.[10]

Public Financial Institutions

Public financial institutions are important in Venezuela. Their financial flows are nearly one-half of total financial flows of the country and this ratio has tended to increase. They are all special-purpose organizations, designed to satisfy a particular need. The four largest in recent years are the Worker's Bank, which funds low-cost housing by constructing buildings and then either selling or renting the units; the Venezuelan Development Corporation, which finances commerce and industry; the Agriculture and Livestock Bank; and the Guayana Corporation. In 1974 the government established others, including the Venezuelan Investment Fund, agricultural and industrial development funds, and an export promotion fund.

These public organizations remedy in part the deficiency in long-term lending of the private financial sector. They use about two-thirds of their resources to buy stock in enterprises, make long-term loans, or buy fixed assets, whereas the private sector uses only about one-third of its resources for such purposes. The public and private financial institutions are not competitive but rather they tend to complement one another. Indeed, the government founded most of the institutions in order to cover a deficiency in private finance.

The government established some of the organizations even before 1950 but it was not until the 1960s that they experienced rapid growth. The Venzuelan Development Corporation, for example, formed in 1946, operated on a limited scale for more than a decade. Between 1946 and 1957 the Corporation loaned only Bs74 million to the private sector. By

1964 its net investments were Bs1,289 million, and at the beginning of 1972 it held Bs1,813 million in stocks and bonds and had made loans of Bs689 million.

The Agricultural and Livestock Bank expanded rapidly in its first years after its establishment in 1943. It fell into desuetude, however, during the 1948–1958 dictatorship. Since its reorganization in 1958, it has become the most important financial institution in Venezuelan agriculture, making primarily short-term working capital and inventory loans. Its operations in the 1960s were greater than in the 1950s, but from the mid-1960s to the end of the decade expansion was modest. In the 1970s the government once again pumped resources into the bank which focused its loans on the campesino sector, and the government formed another bank, the Agricultural Development Bank, to serve commercial farms and livestock interests.

Most public financial institutions are lending, not grant, institutions. They do not, however, have the same stiff banking criteria in use in the private sector and for that reason losses, made good by the government, are not uncommon. They have played an increasingly important role since 1960, not only adding to the total financial resources reaching the private sector, but also in channelling funds into activities which private institutions do not fund.

Capital Markets

Formal capital markets date from 1947 when the Caracas Chamber of Commerce formed the Caracas capital market. In 1958 the Miranda Chamber of Commerce, just across the Federal District line but still in Caracas, formed another capital market. Both operate inefficiently and on a small scale. Although the volume of transactions moving through these markets has increased, they serve primarily the government, not private enterprise, and have not prospered. The issues are the same on both markets but the volume of business is not enough to keep even one market busy. A new Capital Markets law in 1973 offers some promise that the capital market can expand and assist more effectively in private finance. To date, however, the capital market is the weak link in Venezuelan private finance.[11]

The Caracas company is an anonymous company, owned by an insurance company and other stockholders which are not members of the exchange. Dissatisfied members of the exchange defected and formed the Miranda market where the members own the market. Despite its later start, the Miranda market, in the early 1970s, was handling a volume of transactions close to that of the Caracas market. Both markets together handled transactions valued at Bs1,369 million in 1973, a figure less than 9 percent of the loans and investments of the commercial banks.

In the first few years of its existence the Caracas market handled more stock than bond transactions. That changed in 1950, when government securities jumped into predominance, a place they relinquished only in 1953, 1957 and 1958, years of market prosperity. Since 1958 government bonds have dominated both markets, although private bonds increased in 1964 and later. In 1973 only 5 percent of the transactions were for stocks and 45 percent was for government bonds. The two other components of the markets are private bonds (24 percent of transactions) and foreign exchange (27 percent). This general configuration of market activity has prevailed since the early 1960s.

In 1957, during the oil boom, the capital markets appeared as though they might become a significant element in the Venezuelan financial system. Stock trading surged forward in 1957, stock prices continued high and rising, and private companies placed more bonds through the market. In 1958 the markets continued to prosper and the Miranda market was formed. Oil prosperity and new concession money stimulated the markets; industrial, commercial and construction markets were at all time highs. The boom period, however, contained a number of financial irregularities which damaged the prestige of the markets, both for potential customers and for enterprises which might list their stocks.

The market prosperity did not last. The new government, with different policies, injected economic uncertainty into the market. Still, the momentum of the market carried them through 1958, with some decline in stock trading but a large increase in trading in government bonds. Stock prices advanced slightly. In the face of the gathering recession, however, the market gave way. Then the irregularities came to light, and in 1960 the market collapsed and did not fully recover for a decade. In 1971 stock trading was still only one-third that of 1957. The market came to deal primarily in government bond issues, but with some increase in private bonds. All market price indices declined in the early 1960s, with recovery late in the decade. In the early 1970s the market was prospering again, but without changing its composition.

The Central Bank, beginning in the early 1960s, intervened actively in the capital markets, buying and selling government securities primarily. The Bank's purpose was principally to support the government bond market. This generated a conflict, since Bank purchases of bonds to support their price tended to feed additional reserves into commercial banks which could then expand. Bank withdrawal would permit bond prices to fall, a result not acceptable to the government. Hence the Central

Bank, open market operations were not to regulate credit but to support the government. The capital markets, however, were sufficiently small, and monetary policy subject to so many other limitations, that the loss of open market operations was not serious.

Beginning in 1960 the goverment authorized the stock markets to deal in foreign exchange and such trading has become a significant component of market operations. Through the years about one-fourth of market activities have been in foreign money. Such trading, however, is not a necessary element of market operations since the commercial banks and the Central Bank are capable of handling all the necessary foreign exchange activity.

The traditional way in which Venezuelan enterprises raise money inhibits market activity. Most enterprises belong to a family, and should additional funds be needed, the borrowing is within the family. If family resources are insufficient, then the firm seeks out its friends and issues stock or obligations privately. The friend may be another enterprise, a wealthy individual, or a bank, or even in some cases, foreign sources. Most firms see little advantage in listing on the capital markets, especially in view of the financial statements and disclosure requirements necessary for such a listing.

The capital market has not achieved the scope and character of operations necessary to make a significant contribution to private financial development. In part, this is a reflection of the faulty organization of the market, as well as the failure of domestic enterprises to list issues and rely on the market for financing. In addition, stock and bond purchases have not been a popular form of savings by individuals. In the late 1960s and 1970s the market was the subject of intense study, and in 1973 a new law attempted to provide additional incentives both to buyers and companies listing stocks. The advantages were primarily for the buyer and companies did not rush to list new issues. The government continues to consider new measures to fortify the capital market.

NOTES

1. Data on Venezuelan money, banking and finance are in the annual *Informe Económico* of the Banco Central de Venezuela. Data before 1940 are in the *Anuario Estadístico*.

2. Mercedes Carlota de Pardo. *Monedas Venezolanas*, 2a edición, Caracas: Banco Central de Venezuela, 1973, is a detailed account of Venezuelan coinage and money. See also Tomás Enrique Carrillo Batalla. *Moneda, Crédito, y Banca en Venezuela*, Caracas: Banco Central de Venezuela, 1964.

3. Rafael J. Crazut. *El Banco Central de Venezuela: Notas sobre la Historia y Evolución del Instituto 1940–1970*, Caracas: Banco Central de Venezuela, 1970.

4. *Resumen*, 10 August 1975, pp. 24–25; 7 September 1975, p. 42; 19 October 1975, pp. 32–33; 4 January 1975, pp. 44–51.

5. J. M. Herrera Mendoza. *Historia de las Actuaciones del Banco Central de Venezuela*, Caracas: El Cojo, 1960. This volume covers some of the controversy over central banking and the early years of the Banco Central.

6. Alfredo Machado Gómez. *Crisis y Recuperación*, Caracas: Banco Central de Venezuela, 1972. This book covers the recesssion period and was written by a former president of the Banco Central. In his collection of articles, *Ensayos sobre Economía*, Caracas: Banco Central de Venezuela, 1965, Ernesto Peltzer, long-time high level advisor to the Banco Central, covers the early years as well as the recession.

7. Tomás Enrique Carrillo Batalla. *Moneda y Banca en Venezuela*, a special issue of the *Revista de Hacienda*, No. 37, December 1959, discusses the early years of banking in Venezuela.

8. José M. Tejero y Henry Gómez. *La Banca Comercial en Venezuela*, Caracas: Universidad Catálica de Andrés Bellow, 1967. Domingo Alberto Rangel in his *La Oligarquía del Dinero*, Caracas: Editorial Fuentes, 1972, provides a graphic picture of the interrelationships among banks, industry and the economic oligarchy.

9. Rafael Crazut. *El Banco Central de Venezuela*, 1970, pp. 153–187, discusses in detail rediscounting and the scope and limitations of rediscounting policy of the Banco Central.

10. James A. Hanson. "Cycles, Growth, and Structural Change in Venezuela, 1950–1974," a paper presented a the meeting of the Latin American Studies Association, November 1974.

11. Deltec Panamérica. *El Mercado de Capitales en Venezuela*, México: Centro de Estudios Monetarios Latinoamericanos, 1968, pp. 87–128.

Chapter X

International Trade and Finance

International trade is and always has been the lifeblood of the Venezuelan economy. Whatever dynamism the economy has attained, whatever level of income and welfare it has achieved, whatever setbacks and losses it has suffered have come initially through international trade. Its principal export for half a century has been petroleum, and in its market Venezuela had no influence until 1973. For decades imports stunted domestic agriculture and industry. Foreign investment removed decision-making, for much of the economy, from Venezuelan hands.

In 1935 exports were Bs712 million; in 1974 they were Bs45,988. In both years, and all the years in between, petroleum dominated exports. Between 1935 and 1950 exports went up 5.5 times, with the only reverse in the exports attributable to German submarine warfare in 1942 and 1943. From war's end through 1948 exports more than tripled. Exports had increased only modestly but then declined to the 1936 level in the mid-war of 1943. Then exports took off, increasing nearly 13 times in the succeeding five years. The busy industrial world pulled Venezuela along in these years and the economy became addicted to imports.

International trade and investment carried Venezuela to the dizzying heights of prosperity in 1950 to 1957. Everything seemed to go right. Exports were high and rising. Venezuela's principal competitors were in political crises and their supply impeded. New concessions brought in mountains of money and stimulated foreign investment. Then, at the very height of prosperity, everything went wrong. Export prices fell, exports fell, foreign investors lost confidence, the government changed hands and political turmoil ensued, and the economy went into recession.

The sagging world petroleum market reduced Venezuelan exports, and for more then a decade Venezuela's trade was in the doldrums. The new government, in putting hard pressure on the petroleum companies, encouraged them to stabilize exports and expand elsewhere, and reduce their investment. For Venezuela trade was a sputtering engine of growth in the 1960s. Only when the storm clouds of the gathering energy shortage were unmistakable in the early 1970s did exports begin to move ahead. Then in the fall of 1973 the price of petroleum exploded. Exports in 1974 were 3.8 times their value in 1972 although volume fell. Once again petroleum and trade had carried Venezuela to the top of the mountain.

Exports

Venezuelan exports increased from 1950 to 1957, stagnated from 1957 to 1970, and shot up rapidly from 1970 to 1974. The value of exports increased from $1,155 million in 1950 to $2,751 million in 1957, a steady upward trend, with the largest percentage increase in 1957, of 23 percent. The next two years exports fell to 84 percent of the 1957 figure. For eleven more years exports varied around this low level, not achieving the 1957 exports again until 1970 when they were $2,756 million. The stagnation came to an end in 1971 when exports advanced 23 percent, and in 1972 with another 28 percent increase. These years, however, were modest preludes to the jump of 48 percent in 1973, and 86 percent in 1974 to $15,207 million. The statistical record of the four earlier years and the 17 years leading to 1974 are identical: exports increased 5.5 times.[1]

Petroleum has consistently been 90 percent or more of total exports. The proportion was higher—95 percent—until the mid-1950s and in the 1970s. Crude oil is the most important single product of exports. In 1950 Venezuela exported 1,240,000 barrels per day. Crude exports increased to 2,074,000 barrels per day in 1957, and then fell off to 1961 when they were about the same as in 1957. In the rest of the 1960s crude exports edged up, peaking at 2,435,000 barrels per day in 1970. Since 1970 they declined, and were down to 1,772,000 barrels per day in 1974, with 1975 lower still.

Refined products exports were only 184,000 barrels per day in 1950 but experienced rapid and continuous increases until reaching 921,000 barrels per day in 1965. Since that time, these exports have not increased a great deal—in the 1970s running about 1,000,000 barrels per day. The highest exports were in 1974, at 1,074,00 barrels per day. Crude and refined products reached their highest exports in 1970, at 3,470,000 barrels per day. By 1974 these were down to 2,740,000 barrels per day, and in 1975 even lower.

Crude oil prices (which are indicative also of refined product prices) rose in the 1950s to 1957 and then fell until 1961, followed by a gentle drift downward to 1970. Since 1970, the price trend has been strongly upward. The crude price was $2.12 per barrel in 1950, $2.59 in 1957. By 1961 it was back down to $2.08. Then the slide began, reaching a low, finally, of $1.78 in 1970. In 1971 it was up 28 percent to $2.27, rose to $2.43 in 1972, and $3.56 in 1973. *International Financial Statistics* shows an increase of 381 percent over 1971 by the first quarter of 1974, and 501 percent by the second quarter of 1975.[2]

The causes of the behavior of Venezuelan exports and export prices reside in the world petroleum market, in the economic growth of the industrialized countries, and in the petroleum policies of the principal exporting countries, including Venezuela. In the early 1950s the Korean war, the cold war and high international tensions, as well as Venezuelan petroleum policies favorable to the companies buoyed up its exports. European and Japanese economic recovery and strong economic growth in the United States also helped. The continuous Middle Eastern turmoil, erupting in war and the Suez closure favored Venezuela through 1957.[3]

The huge supplies of the Middle East began to have an effect, despite the Arab-Israeli dispute. Middle Eastern oil was cheaper and the transport cost advantage to Venezuela disappeared with the supertanker. Protectionist forces grew stronger in the United States and first succeeded in imposing voluntary imports quotas. Later, in 1959 the controls became mandatory, and with Venezuela outside the perferred area, it lost out in the United States market. International tensions receded. The petroleum shortage disappeared and a glut seemed to be in the offing. The new Venezuelan government adopted a much tougher petroleum policy and through the decade of the 1960s maintained pressure on the oil companies, which responded by holding the line in Venezuela and expanding in the Middle East and Africa.

The appetite of the industrial countries for petroleum all through the decade of the 1960s was voracious. Petroleum fueled industry and kept millions of cars and trucks on the road. House heating burned up millions of barrels of oil and the burgeoning petrochemical industry drained supplies further. At the same time the petroleum exporting countries were achieving more and more control over their own petroleum sectors and putting pressure on prices. The Arab embargo of 1973 restricted supply and gave the Organization of Petroleum Exporting Countries the opportunity to take price setting away from the oil companies.

Iron ore exports were of no significance until the mid-1950s when they superceded coffee as the second most important export. An uninterrupted upward trend prevailed from 1953 to 1960 when iron ore exports were nearly 7 percent to total exports. In the 1960s they declined and stabilized, but in the 1970s they climbed again near to the 1960 level, partly as a result of increases in price. Agricultural exports, notably coffee and cacao, recently joined by rice, have only a marginal role, and industrial and manufacturing exports almost do not exist.

Imports

Venezuelan imports contain three significant elements: merchandise imports (one-half or more); financial services (one-quarter to one-third); and transportation and other services (one-sixth). These ratios vary from year to year and merchandise imports have tended to encroach on service imports. Merchandise as well as total imports tripled between 1950 and 1957, reaching a level which was not attained again until 1970. The increase in merchandise imports was 51 percent in 1957 alone. Service imports also reached a peak in that year, not regained until 1968. The 1950s was a heady time for imports.

Beginning in 1958 imports in all categories began to go down, slowly in 1958 and 1959 and then a precipitous drop of 25 percent in 1960. Merchandise imports declined to 55 percent of the 1957 high by 1963, a pattern also followed by non-financial services. Financial services began to go up again in 1961 and all imports jumped by 21 percent in 1963. But thereafter in the 1960s, imports flattened out for four years. In 1968 there was a 16 percent increase, but in 1969 and 1970 imports were again stable. In 1971 imports began to move ahead again. Between 1970 and 1974 merchandise imports more than doubled. Service imports more than doubled from 1970 to 1973, and in 1974 alone went up 2.4 times. The 1960s were slow years for imports but with the return of petroleum prosperity in the 1970s imports flooded in.

Bear in mind, however, that much of the payments for service imports relate to exports. Income paid to the owners of the petroleum companies and for transport services—which is one-half of non-financial services—are in fact a part of the cost of exporting petroleum. Without these payments the companies would not work and the oil would not move. Of the order of 40 percent of imports are thus necessary in order to obtain the remaining 60 percent of imported goods.

The ups and downs of imports reflect, somewhat imperfectly, export proceeds. International reserves were never sufficient, until 1974, to permit unbridled imports without exports to match. What happened in 1974 confirms the 1957 tendency to splurge on imports when reserves are high. The downward movement in 1958 and after, reflects the recession, export stagnation, as well as the determined import restriction policy and exchange controls, although these were not sufficiently stringent to establish a permanent current account surplus.

The merchandise balance was solidly positive, the service balance solidly negative, and the current account has been positive for the past twenty-five years, but with many years in which modest deficits offset modest surpluses. The high imports of 1956–1959 resulted in deficits which wiped out the surpluses of 1950–1955, as well as those of 1960 and 1961. Deficits turned up again when imports accelerated in 1968 and persisted for three years. The windfall surpluses of 1973 and 1974, however, generated a surplus of $8.7 billion, 71 percent of which came in 1974. Imports and exports were thus approximately balanced over the last twenty-five years, except for 1974.

Direction of Trade

The principal trading partner of Venezuela has been in the United States. The United States buys crude and refined products, and Venezuela is an important element in covering the United States fuel deficit. Even the crude shipped to the refineries in the Netherland Antilles from Venezuela moves into the North American market after refining at Aruba and Curacao. During the 1950s the United States share of Venezuelan exports tended to decline as Middle Eastern oil gained a foothold in that market. In 1960, 50 percent of Venezuelan exports went to the United States, another 9 percent went to Canada. Latin America took 22 percent and Europe 19 percent.

In the decade of the 1960s the North American share declined. In 1969 only 42 percent of Venezuelan exports went to the United States market. On the other hand, the Canadian share increased to 13 percent. The Latin American share increased slightly to 23 percent. The Antilles refined products continued to feed into the North American market. The purchases by Argentina and Chile increased. Europe's share of exports remained the same in the 1960s, with England, West Germany, Spain, Italy, France and Holland as the main buyers. Japan, which made almost no purchases in 1960, bought increasing amounts and in 1969 had 2 percent of Venezuelan exports.

During the decade a triangular trade among the United States, Canada and Venezuela developed. Canada increasingly supplied the United States market from its interior Alberta Field. Hence, Venezuela lost out, in part, in the growth of the United States market. On the other hand, the Canadian East Coast was increasingly supplied by Venezuela. The Western Hemisphere share of Venezuelan exports remained the same; the United States and Canada, in part, traded positions. In the 1970s the United States regained its position. Including Antilles refined products, the United States took 49 percent of Venezuela's exports in 1972 and 53 percent in 1973, up from 43 percent in 1969. The increased United States share was at the expense of Western Europe (whose share was nearly cut in half) and Canada.

The United States has also been the dominant supplier of imports to Venezuela, providing about one-half of what Venezuela buys. In 1960 the United States share was 52 percent, in 1969 it declined to 50 percent. For the 1969–1972 period, the United States share was 47 percent. Canada's share, although small, increased. West Germany is the second largest import supplier, about 10 percent of the total. Japan ranks third, and its supplies to Venezuela are growing faster than that of any other country; its share reached more than 8 percent in 1969–1972. While the United States dominates Venezuelan imports, its influence is loosening slightly.

Following the dominant influence of petroleum prices, the export prices of Venezuela (before the bonanza of the 1970s) reached their peak in 1957—26 percent higher than in 1950 and 30 percent higher than in 1969. The rising trend was modest from 1950 through 1956, but in 1957 export prices leaped 10 percent. But prices receded almost as quickly; by 1960 they had fallen off 20 percent from their peak and were back to the 1952 level. In the 1960s they continued a downward slide and by 1969 were 85 percent of the 1960 level. Between 1972 and 1975 export prices quintupled under the impact of restricted supply and rising demand.

Rising, but less pronounced, trends prevailed for import prices. In the 1950s they advanced 63 percent. The 1960s started off with modest price increases but in 1962 they went up 5 percent, followed by declines for 1963 and 1964. Jagged and trendless price changes in imports prevailed from 1965 through 1970, with prices in the latter year 6 percent above those of 1965. Between 1971 and 1974 import prices advanced by 34 percent, mostly in 1973 and 1974.

The derived trend for the net terms of trade, that is, export prices divided by import prices, was almost continuously downward until the effects of the price increases of 1973 brought export prices back into line with import price increases over more than two decades. Taking 1959 import and export prices as the base, the most favorable net terms of trade experienced by Venezuela before 1974 was in 1950, 49 percent higher than in 1959. The 1950s were up and down years, and even 1957, the year of the highest export prices, showed terms of trade 9 percent inferior to 1950 and lower than in previous years. The largest drop came in 1959, when the terms of trade plummeted by 28 points. In the 1960s the terms of trade continued to deteriorate but less severely than in the previous decade, declining by 1969 to 78 percent of 1959.

The quantum of exports—a measure of the physical volume of exports—were on a rising trend until 1970 when the peak petroleum exports occurred. Export quantum grew 151 percent between 1950 and 1969, increased modestly in 1970, and has declined in the 1970s. Import quantum has been more irregular, rising in the 1950s to 1956, then a giant leap of 48 percent in 1957 alone, followed by reductions to 1963. Import quantum subsequently increased erratically. The gross terms of trade—the amount of exports per unit of imports—had no trend in the 1950s until beginning in 1958 and lasting through 1963, increasing amounts of exports were needed for each unit of imports. Later in the 1960s the gross terms of trade declined and, of course, in the 1970s dropped sharply to well below the 1950–1969 levels.

The capacity to import—the net terms of trade times the quantum of exports—increased regularly in the 1950–1957 period as the increasing export quantum offset the declining terms of trade. The sharp drop of 1958 and 1959 ushered in a decade, however, when the capacity to import moved irregularly around the 1959 value, which was 29 percent below the 1957 high. In the 1970s the improvement in the terms of trade was so great that it more than offset the declining quantum of exports, and in 1973 and 1974 especially the capacity to import improved.

The outstanding feature of price movements was the increase, decline, and rise again of export prices, coupled with a steady rise in import prices. Export volumes moved up, stagnated, and declined. The greatest benefits of trade to the Venezuelan economy came, therefore, in the 1950s until 1957, with moderately high prices and rising volume, and in the 1970s beginning in 1972 when prices began to move ahead strongly despite declining volume. From 1958 to early 1970s trade made a contribution but was not pushing the economy along.

The Venezuelan terms of trade experience gives small comfort to the proposition that the terms of trade of countries of the periphery, the primary producing countries, tend to decline in a secular fashion. Twice in less than two decades petroleum prices increased, once modestly under the auspices of an oligopolistic industry, the other time grotesquely with price under the control of an intergovernmental cartel. And even the most optimistic petroleum user accepts the 1974–1975 price as the basis for long-run increases, with declines, if any, small and temporary. Center-periphery theorists must seek elsewhere for confirmation of long-run deterioration in the terms of trade for primary producing countries. Indeed, the events to 1973–1975 may well fatally undermine the hypothesis, at least for exporters of nonrenewable resources.[4]

Trade Policy

Government policy has been to encourage the increase of exports of all kinds, in the hopes of earning more foreign exhange, and to limit imports to conserve foreign exchange and to protect domestic industry and agriculture. Venezuela has pursued a nationalistic and mercantilistic policy. Some policies acted directly on domestic producers to encourage export or restrict imports. Others impeded imports at the customs house. The government did not regard taxes on the petroleum sector as a part of trade policy, but rather as fiscal policy. Other trade policies have included participation in international organizations to influence imports and exports, although Venezuela has not been a member of the General Agreement of Trade and Tariffs.

The traditional goal of export policy has been to maintain and increase the amount and share of Venezuelan petroleum and mining exports and to increase export prices. The change in petroleum policy in the 1960s, however, signalled a reduced emphasis on increasing export volume of petroleum, while increasing export values through rising prices. Earlier the government had considered that it could not influence price and that the only way to earn more foreign exchange, except fortuitously, was by exporting more volume. The decision to participate in founding the Organization of Petroleum Exporting Countries in 1960, and supporting it consistently reflects this new policy, as does the reduction on the volume of petroleum exports since 1970, especially in 1974 and 1975.

The earlier eagerness to increase petroleum exports led Venezuela into the trade agreement with the United States in 1939 and to its renewal in 1952. It provided a tariff preference for Venezuelan oil at the cost of reductions of import duties on manufactures, and binding these low rates. The United States, however, in the mid-1950s installed a voluntary quota system which did not favor Venezuela, and the 1959 mandatory quotas were also not in Venezuela's interests.[5] Venezuela protested mightily without avail, violated it in part in the early 1960s in order to protect itself, and in 1972 abrogated the treaty, not only because oil exports had ceased to be as important a problem, but also because the agreement still bound rates and impeded other international polices which Venezuela wished to pursue, most notably entry into the Andean Group.

Venezuela has endeavored, without much success, to diversify its exports. Within the petroleum sector, its policy has been determined and consistent: instead of shipping crude, Venezuela's policy called for exporting increasing quantities of refined products. For other sectors of the economy the policy of export diversification has primarily been wishful thinking, as the Venezuelan cost structure renders non-competitive both its manufacturing and agricultural products. In a few instances, such as the rice surplus in the early 1970s, a new export turned up but no important broadening of the export base has occurred. Venezuela is using a small part of its petroleum bonanza to seek out and subsidize and to offer tax advantages to non-traditional exports, but the policy of changing the structure of exports has been unsuccessful.

More out of concern for hemispheric and regional solidarity than the real potential for export promotion, Venezuela joined the Latin American Free Trade Association in 1966 and the Andean Group in 1972. Domestic industrial interests resisted both, and reluctantly acceded, although fearful of low-cost imports and incapable of accepting the export challenge. In the 1970s Venezuela has reached out with new financial ties in the Caribbean area and Central and South America, partly in hopes of promoting export markets. Venezuela has spoken strongly in favor of the export interests of primary producing countries and expressed outrage against the United States' denial of preferences to OPEC countries.

Import policy consists primarily of import duties. Until the mid-1940s these served a revenue function but have since been protective. The protection was not rigorous and the trade agreement with the United States bargained away many duties which might have promoted industrial development. The 1959 government, determined to foment industry and agriculture, found loopholes which permitted greater protection through higher duties, quotas and other non-tariff barriers. In 1972 a new system of *ad valorem* duties replaced the old specific duties, and new and higher tariff schedules took effect.[6]

Import duties are highest for consumer goods which are or can be produced in Venezuela. For example, the duty on new cars is 300 percent and similar duties prevail for consumer durables and other products. Capital goods face moderate duties and raw materials the lowest duties. On raw materials and capital goods the government often exonerates the duties, especially when these imports serve a development objective of interest to the government. In addition to import duties, Venezuela also employs non-tariff barriers, such as quotas, licensing, "buy Venezuela" rules, consular and documentation fees, import prohibitions and other measures, some of which serve protective purposes. Domestic prices, however, are so high, and the preference for foreign goods so strong, that in spite of protection imported goods are ubiquitous in Venezuela.

Balance of Payments

Short-term private capital movements into and out of Venezuela were modest during most of the 1950–1969 period. For the twenty years, $833 million left, $349 million entered. The negative balance is largely attributable to three years, 1961–1963 when $660 million in short-term private capital fled the country. It was an era of great economic and political uncertainty, with terrorism rampant and doubt about the stability of the new constitutional regime. The harder line on petroleum and the introduction of economic planning also abetted the capital flight, as did the recession, high unemployment, and stalled growth. Leaving those three years out, the short-term private capital account, however, was in the black. Between 1969 and 1974 a small amount of capital flowed into the country.

The government short-term capital account was not active and was slightly positive, except for 1958 and 1959 when it made payments abroad of $619 million. These were debt payments for obligations incurred by the previous regime and the government hastened to pay them in order to avoid international pressure which might add to uncertainties about the new government. Except for these two years, there was a small inflow most of the years.[7]

Influenced heavily by the two abnormal periods, 1958–1959 and 1961–1963, in which $1,333 million in capital moved out, the short-term capital account shows a net outflow of $1,092 million between 1950 and 1969, and a net inflow of $143 million from 1970 to 1974. Subtracting those five years—1958–1959 and 1961–1963—there was a small net inflow. However, the timing of the capital flights, in relation to the change in government and the recession, makes them important in policy formation, most notably in new import restrictions, the imposition of exchange controls, and the devaluation of the bolivar.

The long-term capital account consists of the petroleum, iron ore, and rest of the economy sectors. In all, Venezuelans invested $1,009 million abroad and foreigners invested $2,354 million in Venezuela between 1950 and 1969. In two years, 1956 and 1957, at the time of the Suez crisis, substantial investment occurred. The net foreign long-term capital inflow on the petroleum account was modest, only $209 from 1950 to 1959. The figure is small because of disinvestment during the 1960s; although gross investment was $1,589 million, of which nearly three-fourths was in the two concession years. Small amounts of net investment began again in the 1970s, but by that time it was clear to the companies that they had only a short future in Venezuela.

In every year of the 1960s except the last two, the companies disinvested, cumulating in 1960–1967 to $1,199 million, offsetting almost exactly the 1956–1957 investment, attributable to new concessions and the Middle East crisis. The reverse flows of the 1960s result from the failure to the companies to obtain more concessions, to more attractive opportunities in the Middle East and Africa, and to tax and regulatory pressure from the government. In 1968 and 1969 the companies built desulphurication additions to their refineries under government guarantees, and the investment of the 1970s resulted from company-government negotiations.[8]

Iron ore was a relatively small account and the industry removed about the same amount that it invested. Since the net capital movement in petroleum and iron ore was a modest $215 million between 1950 and 1969, the overall net inflow of $1,345 million was in the rest of the economy, that is, investment in industry, commerce, services and banking. In only three years, 1951, 1959 and 1961 was there a removal of long-term capital, although the inflow declined sharply in 1958. From 1961 to 1970 foreigners made investments in Venezuela to a greater extant than Venezuelans made investments abroad. In 1970 there was a small outflow, and in 1973 and 1974, $251 million in long-term capital left Venezuela.

The total capital account shows a small net inflow of capital between 1950 and 1974, $521 million, with large year-to-year variations. The strong inflows of 1956 and 1957 show up clearly, as do the following six years of economic and political problems, debt repayment, short-term capital flight, and petroleum disinvestment. The two years show an inflow of $1,498 million and the succeeding six an outflow of $1,917 million. From 1958 to 1967 only one year, 1965, displayed an inflow, a small one. In the 1970s through 1972 a small net capital inflow occurred, but in 1973 and 1974, when reversion and nationalization was in the wind, capital left, especially in 1974.

Adding the current to the capital account alters the trends. The current account deficits mute the capital inflows of 1956 and 1957 but the years up to 1958 show an accumulated surplus on current and capital accounts of $1,247 million, two-thirds of which dissipated in 1958 and 1959. The surplus began to mount again, reaching $1,325 million in 1969. In 1970 and the years following, the combined accounts are in surplus each year with a huge $4,080 million surplus in 1974 alone. Errors and omissions are negative for most years (in most years of small magnitude) but in 1973 and 1974 of increasing importance.

The petroleum sector contributed $587 million to the balance of payments in 1950, and each year it increased during the decade with the peak contribution of $1,874 million in 1959. During the 1960s, however, the contribution was about $1,500 million each year, and it did not increase until the 1970s. The iron ore contribution was small and variable; in the 1960s it was twice that of the 1950s. The negative components in the balance of payments came from the rest of the economy. These also grew in the 1950s, reaching their highest level in 1959 and then falling and levelling off in the 1960s, then growing again in the 1970s.

International Reserves

Venezuela's international reserve position has been strong throughout the last twenty-five years. These reserves consist of a small amount of gold, deposits in foreign banks (primarily in dollars), a credit position with the International Monetary Fund, and holdings of bills, U.S. Treasury letters and certificates of deposits, and during the 1970s, special drawings rights. Except for 1956–1958 when reserves piled up rapidly, and in 1963 when imports dropped substantially, most foreign bank balances were in time deposits. Most of the time, Venezuela has held its own gold.

In most years there was an accumulation of reserves, with two outstanding aberrations. From 1950 to 1955 a slow accumulation occurred, amounting in total to $197 million net. Then in 1956 and 1957 reserves shot up, partly because of the new concessions, and partly because of increased petroleum exports at high prices. In these two years, $857 million additional reserves brought in 1957 reserves to an all-time high up to that point, of $1,396 million. The reversal of previous trends, however, then led to rapid decumulation in 1958–1960, a period in which $791 million in reserves were lost. Hence in 1961, reserves were only slightly more than in 1955.

From 1963 through 1969 once again reserves accumulated, with only a minor loss in 1966. The total net accumulation was $418 million but no trend appeared. In 1969 reserves stood at $939 million. In 1970 a small amount of additional reserves came in, but in 1971 the accumulation pushed reserves over the previous all-time high in 1957, reaching $1,522 million. Reserves continued to build with the largest annual increase in 1975, a jump of 270 percent over the previous year, to $6,513 million. In October 1975, reserves were $8,781 million, equal to merchandise imports in 1972, 1973 and 1974 combined.

The Central Bank captures nearly all international reserves. Commercial banks characteristically hold only 1 or 2 percent as a convenience to the public and have ready access to the Central Bank's reserve fund when needed. The Central Bank also holds the country's gold. The ratio of gold to total reserves has varied greatly. In the early 1950s reserves were primarily gold, but during the decade the gold coefficient declined. It went up significantly in 1959 when bank balances instead of gold paid off debts, but during the 1960s the gold-reserve ratio declined again, with a constant gold stock of a little more than $400 million. In the 1970s the Bank made some gold purchases and at the end of 1974 the Bank held $488 million, a small fraction of total reserves.

Exchange and Exchange Rate Policy

The Central Bank, since its inception, has managed all foreign exchange. The petroleum companies turn their foreign exchange earnings, primarily dollars, over to the Bank and receive bolívar deposits. Iron ore companies do also, as do the producers of coffee and cacao. The commercial banks maintain small foreign exchange balances for the convenience of business and the public, and when necessary replenish their supply at the Central Bank. The bolívar is freely convertible into any currency at the official exchange rate of Bs4.285 to the dollar and the quoted daily rates for other currencies.

At the end of 1960 Venezuela introduced selective exchange controls in order to dampen the large short-term capital flight. The Central Bank managed the controls for the government. They were reasonably effective; but probably more effective in stemming the capital flight was the stability of the government, the failure of terrorism, the return of confidence, and the end of the recession. Exchange controls ended in January 1964, and at the same time the government eliminated most of the system of multiple exchange rates and devalued the bolívar.[9]

Venezuela joined the International Monetary Fund in 1945, but instead of a single parity rate the government paid the oil companies Bs3.09 to the dollar, and cacao and coffee producers substantially more. The Bank sold foreign exchange at a nearly uniform rate of Bs3.35 to the dollar. In 1964 the government established Bs4.40 to the dollar as the petroleum rate and $4.50 as the market rate for nearly all other transactions, maintaining only minor differentials for the coffee and cacao producers.[10] After the dollar devaluation in 1971, which automatically devalued the bolívar for non-dollar pegged currencies, the government revalued the bolívar to Bs4.30 for petroleum and Bs4.40 for other transactions. When the dollar devalued again in 1973, the bolívar again was revalued, to Bs4.20 for petroleum and Bs4.30 for other transactions. The official IMF parity rate is Bs4.285 equals $1.

In 1974 and 1975, with large and mounting international reserves and inflationary pressures at home, Venezuela considered and rejected revaluation. With nationalization, most exchange earnings go directly to the government and revaluation would have no effect which could not be achieved by domestic monetary controls or sequestering reserves as in the Venezuelan Investment Fund. Revaluation would make the necessary imports, already a source of domestic inflation, even more expensive. Import duties and similar policies, becoming increasingly more strict, can hold down the luxury imports.

International Indebtedness

Venezuela has had a terror of international indebtedness. Shortly after the turn of the century, European navies bombarded Venezuela and bottled up its tiny navy in order to collect debts. General Juan Vicente Gómez learned the lesson well, paid off all foreign debt, and refused to borrow. The Gómez no-debt tradition lived after him. Except for current account obligations, Venezuela had no foreign debt until 1958. Since then the debt has crept up, but under the strict surveillance of Congress which has to approve all foreign borrowing.

Despite the dislike for debt, however, beginning in 1958 it was necessary in order to pay obligations piled up by the dictator. Direct debt jumped from near zero to Bs688 million in 1960, although the indirect debt, generated by government agencies and guaranteed by the government, had begun two years earlier. Thereafter, both direct and indirect foreign debt went down until 1966 when the indirect debt began to increase beyond the 1958 level.

In 1970 both kinds of debt, each about the same amount, summed to Bs2,932 million, slightly more than one-half of the total public debt. Since 1970 the debt has continued to increase slowly. At the end of 1974 the foreign debt was Bs3,293 million, but by this time the direct debt was only Bs879 million of the total, and the internal debt was more than the foreign debt. Of the total foreign debt, 35 percent was due the International Bank for Reconstruction and Development, and another 13 percent to the Interamerican Development Bank. The remainder was scattered among a variety of foreign, private and public lenders.[11]

Recent International Economic Policy

The influx of petroleum revenues since the fall of 1973 has presented Venezuela with the problem of what to do with all the money. To spend it all in economic development and the expansion of government would create intolerable inflationary pressures and lead to waste. The government in 1974 decided to embark on a more active international economic policy which would make use of the new funds until they were needed for domestic development and to fortify its leadership position among less developed countries.

One accelerated activity is in international organizations. Venezuela, by the end of 1974, had committed $540 million to the oil facility of the International Monetary Fund for 1974, and later committed $240 million for 1975. Its reserve position in the Fund in December 1975, was $892 million. It also loaned $500 million to the International Bank for Reconstruction and Development and $500 million to the Interamerican Development Bank, in both cases far in excess of Venezuela's debt to those organizations. Venezuela has urged and obtained an increase in its IMF quota, as well as its quota in other international organizations, and has taken an active role in discussions leading to the reform of the international monetary system in Jamaica in January 1976.

Venezuela has loaned the Central American republics $600 million, to be disbursed over a five-year period, to stabilize coffee prices and permit those countries to buy oil. It has also subscribed to bonds of the Central American Bank for Economic Integration. Other bilateral arrangements include loans to Argentina and Peru and negotiations with other countries, including the Caribbean countries, whose balance of payments have been adversely affected by the increase in the price of oil. In effect, Venezuela now has a foreign aid program of its own whose relative magnitude is many times that maintained by the industrialized countries.[12]

In addition to the bilateral and multilateral arrangements, Venezuela has become more active in international financial markets. The Venezuelan Investment Fund in December 1975, stood at Bs22,500 million, consisting mainly of $4,257 million invested in international capital markets. The money is primarily in short and intermediate-term financial securities. An equal amount, the remainder of Venezuela's international reserves, is also invested abroad in capital markets. These activities simultaneously support the recycling of petroleum earnings back into the international financial system, earn interest for Venezuela, isolate liquidity from the domestic monetary system, and assure the availability of funds for future development projects.

Its new position has made Venezuela a leader and spokesman for less developed countries in international economic policy. It supports policies leading to a redistribution of economic power and wealth among the countries of the world. It has proposed international commodity agreements and other arrangements like OPEC, the indexation of export prices of industrial countries and primary producing countries, monetary arrangements more favorable to the less developed countries, economic integration and coordination of economic policies of less developed countries, and other measures designed to accelerate the growth of the less developed countries of the world.

Summary

International trade sets the pace for the Venezuelan economy. When the petroleum export market prospers, the Venezuelan economy booms; when the market drags, the economy is in recession or grows slowly. In the last four decades trade and the economy have moved in giant waves. By 1935 Venezuelan trade was coming out of the Great Depression, and the war and postwar international prosperity and tensions resulted in the growth of exports, as well as imports into its relatively unprotected domestic market.

The market broke in 1958 and stagnated for more than a decade. The economy responded with recession and slow growth, although Venezuela cut imports and protected the domestic market more adequately. When the market turned again in the 1970s, Venezuelan growth speeded up and the country fortified its reserve position. Through it all, except for a few years, Venezuela maintained a favorable balance of payments position, accumulated reserves, and minimized debt.

The basic character of Venezuelan trade, however, has not changed. Petroleum is still the export. Indeed, if anything, trade in petroleum has a tighter grip of the Venezuelan economy than ever before. Imports, especially of manufactured goods, are still necessary. Even though the domestic market has protection, imports are high and in evidence everywhere in great quantities. New resources in the last few years have permitted a more open and supportive international economic policy. But while all the numbers are larger and the Venezuelan position at this time is more favorable, the structure of trade remains the same.

NOTES

1. All of the basic trade data—imports, exports, prices—are in *La Economía Venezolana en los Ultimos Treinta Años*, 1971: and in the annual *Informe Económico* of the Banco Central de Venezuela; as well as the *Anuario Estadístico*. The Ministry of Development publishes the monthly *Boletín de Comercio Exterior*; and *Estadística Mercantil y Marítima*; and the annual *Compendio del Comercio Exterior*. The Ministry of Mines and Hydrocarbons publishes detailed petroleum trade data in the annual *Petróleo by Otros Datos Estadísticos*.

2. *International Financial Statistics*, Vol. XXVIV, No. 1, January 1976, pp. 408–409.
3. Domingo F. Maza Zavala. *Problemas de la Economía Exterior de Venezuela*, 1962, pp. 224–298.
4. Carlos A. D'Ascoli. *Moneda, Comercio Exterior y Otros Ensayos*, 1967, pp. 71–103.
5. Tomás Enrique Carrillo Batalla. *La Economía del Comercio Internacional de Venezuela*, 1962, pp. 7–32.
6. Orlando Araugo. *Situación Industrial de Venezuela*, 1969, pp. 13–28.
7. Domingo F. Maza Zavala. *Venezuela, Una Economía Dependiente*, 1964, pp. 161–205.
8. Franklin Tugwell. *The Politics of Oil in Venezuela*, 1975, pp. 179–181.
9. Tomás Enrique Carrillo Batalla. *La Economía del Comercio Internacional de Venezuela*, 1962, pp. 88–142; Rafael Crazut. *El Banco Central de Venezuela*, 1970, pp. 133–152.
10. Domingo F. Maza Zavala. *La Internacionalización del Bolívar y la Liquidez Internacional*, Caracas: Ediciones del Cuatricentenario de Caracas, 1967, pp. 104–125.
11. *Informe Económico 1974*, Banco Central de Venezuela, p. A-229.
12. *IMF Survey*, International Monetary Fund, 9 June 1975, pp. 174–176.

Chapter XI

Agriculture

Until the advent of petroleum Venezuela was an agricultural country, producing most of its own food and exporting tropical products, coffee and cacao. Most of the people lived in the countryside, eked out a wretched living on large estates and ranches (the haciendas) and small plots (the conucos). Most agricultural production was in the mountain valleys, but only a small fraction of arable land was in use. The sector had little capital, was technologically primitive, and its productivity was deplorably low.

Coffee and cacao were the source of what little prosperity and growth the sector experienced. These export crops occupied about one-half of the land in use; the remainder was in subsistence farming and crops for domestic marketing. Even before petroleum came along, however, coffee and cacao had passed their zenith. All the land suitable for them was in use, the trees were old and costs were rising. Labor was inefficient and the technology was crude. By the turn of the twentieth century it was clear that not only was Venezuela a poor agricultural country but also that its prospects were dim.

Between the first large-scale petroleum exploitation and 1950, agriculture experienced decline and stagnation. The oil expansion required unskilled manpower. Most of it came from agriculture. Other agricultural workers used the petroleum boom as an excuse to escape to urban commercial and service employment or unemployment. The result was rising agricultural labor costs, abandoned haciendas and conucos, and a great wave of migration to the Lake Maracaibo area. Capital also fled from agriculture into industrial, commercial and service sectors which fed off petroleum. The loss of productive factors and growing imports permitted by petroleum exports limited agricultural expansion.

Agriculture declined as an employer and contributor to gross territorial product. Despite the rhetoric, government policy did not change the land tenure system, did not redistribute land, did not increase land in use, capitalization, or productivity.The 1945–1948 reform government attempted to introduce a more active agricultural policy and some agrarian reform, but the succeeding military regime wiped out the advances and ignored agriculture. Dependence on agricultural imports increased as petroleum exports went up. By 1950 agriculture was not a central sector in the economy, except as an employer and subsistence sector, was only 8 percent of gross territorial product, barely viable, and was a drag on the growth of the economy.

Since 1950, most notably since 1958, agriculture has improved. More vigorous agricultural policies and agrarian reform resulted in growth only slightly less than the economy as a whole. Output per worker increased, more land is in use, and capital (private but much of it based on public credit) has created a small modern commercial agriculture and improved the campesino sector as well. The country has achieved near self-sufficiency in some products and is less dependent on imports. While agriculture is still the sick brother of industry and petroleum, the last quarter century has witnessed the stirrings of life and vitality in agriculture.

Structure and Growth

The value of agricultural production from 1950 to 1969 (in 1957 prices) increased 287 percent, and between 1970 and 1974 (in 1968 prices) by 15 percent, an overall rate of growth of somewhat less than 6 percent. Since 1960 agricultural production has grown more rapidly than has population, but from 1954–1956 and 1969–1971 population growth exceeded agricultural growth and in several years agricultural production declined. Wide swings have characterized the agricultural growth rate.[1] The years of the 1950s were feast or famine. One year, 1953, agricultural growth was 18 percent, but the next year production fell by 8 percent. Then in 1955 growth was back up to more than 11 percent, followed the next year by a 1 percent decline. The key crops in these wide variations were sugar, fruits and vegetables, and rice. Despite the ups and downs, however, growth for the decade was 68 percent.

The 1960s started off swiftly with the second highest growth of the past quarter century, 14 percent. It was attributable in part to the initiation of agricultural reform and to heightened import restrictions. The worsening recession, however, began to take its toll the next year in which almost no growth occurred. For the rest of the decade agricultural growth was more stable, and while the growth rates were not as high as earlier, neither were they as low. For the decade, growth was 50 percent, much below that of the 1950s, but more stable and almost equal to the growth of the economy. In the 1970s agricultural growth has lagged, almost stagnant from 1969 through 1972 but a respectable 7 percent in 1973, and 8 percent in 1974.

Agricultural production has made only a small contribution to gross territorial product, much smaller than in most Latin American countries. In 1957 prices, the agricultural share in 1951 was 8 percent and it has tended to decline slightly. In 1973 and 1974 the share was 7 percent. In between, following the vicissitudes of agricultural and economic growth, the share varied slightly. The tendency toward decline, of course, reflects the slightly slower growth of agriculture relative to the economy as a whole.

Venezuela classifies its agricultural production into four categories: (1) crop production, (2) animal production, (3) fishing, and (4) forestry. Despite its long seacoast, Venezuela has never been a fishing nation. Even so, its small production is more than three times its national consumption in recent years and this sector is one of the significant lines of agricultural exports. Venezuela also has large stands of forests but production is small and the country is a net importer. The two combined are less than 5 percent of agricultural production.

Animal production, on the other hand, has been important historically and of growing significance in recent years. In 1950 animal production was 26 percent of agricultural output, but by 1972 it was 49 percent. Each sub-category within animal production grew more rapidly than agricultural production as a whole. Milk production, for example, increased 4.3 times between 1950 and 1969, while eggs increased a fantastic 21 times. These two products were 35 percent of animal production in 1950 and the relationship has remained stable. Cattle production, (that is, output, not the herd) was 47 percent of animal production, but its relative importance has declined since 1950 as the production of fowls and other animal products has increased more rapidly.

Crop production slipped behind animal production in the 1960s. In 1950 crops were 66 percent of agriculture but by 1972 the ratio was only 42 percent. Crops were also more sensitive to the recession and between 1958 and 1962 grew only one-half as fast as all agriculture. The most important crops in 1972 were corn, sugar cane, coffee and bananas. Coffee and cacao were major crops for two centuries, but production has stagnated for more than twenty-five years. In the early 1960s corn superceded coffee as the most important crop and in the 1960s sugar cane has rivalled coffee.

Corn production increased 2.3 times between 1950 and 1971, but it declined recently until 1974 when it increased by more than one-fourth. Rice production was low in the 1950s and rice was imported. Rapid growth from 1962 to 1969 converted rice into a minor export product, but production has stagnated recently. Cotton production grew rapidly, standing in 1973 at double the production of the early 1960s. Sugar cane production also doubled in the same period. Yucca, a major staple food crop, however, had below average and erratic growth. Potato production has tripled since the early 1950s. Coffee and cacao production has stagnated.

Milk, cattle, eggs, fowl and meat production have all had high growth, most of it concentrated in the 1960s and 1970s. Fowl production was 2 million head in 1950, 16 million in 1959, but 89 million head in 1973. Per capita egg production was only 12 per year in 1950, but by 1973 it was nearly 140 per person per year. Per capita milk production more than doubled.

In summary, agriculture has grown rapidly and changed its structure in the last twenty-five years. Its growth was only slightly less than the growth of the economy as a whole. Cereal production, previously dominant, has given way to relatively more animal production. Within cereal production, the staples, such as corn and yucca, have not fared as well as rice, sugar cane, fibers and fruits and vegetables. Within animal production, the high protein foods, such as eggs, milk, fowl and meat have grown most rapidly of all. Based on domestic production, Venezuelans now have a more balanced agricultural economy than in 1950, and one that is more closely suited to the needs of the population[2]

Imports and Exports

Because agricultural productivity has been so low, the costs of Venezuelan agricultural products have been high; without protection, the sector cannot compete in domestic markets. Almost universally, the prices of domestic products are higher than corresponding prices in other countries and imports. In the 1950s the government afforded agriculture some protection through tariffs and quotas, and by producer subsidies. Even so, imports increased more rapidly than did domestic production. Milk product imports doubled during the decade, as did imports of eggs, sesame, legumes and copra. Wheat became a major import. The basic government trade policy was to increase petroleum exports sufficiently rapidly so that the country could import increasing quantities of food.

In the 1960s the government introduced an import substitution program for agricultural products, cutting imports and expanding the domestic agricultural product market. More protection shielded most products and in some cases quotas or prohibitions brought into existence a market for domestic producers. The country became more self-sufficient in many products. With 1959 as 100, the quantum index for food imports was 59 in 1969. The value of food imports increased 15 percent while the value of all imports increased 90 percent.

By the 1970s Venezuela supplied 85 to 90 percent of its agricultural needs. The most important remaining import, accounting for about one-half of total agricultural imports, was wheat. Venezuela has no land capable of producing wheat at anything near the market price. Another important import is temperate zone fruit. Some meat also is imported, primarily as an intermediate product for the domestic food processing industry.

Venezuela exports the two former staple exports, coffee and cacao, as well as some sugar cane and some fish. Again, high Venezuelan costs prevent the country from penetrating foreign markets. It is probable that even some of the coffee exports are contraband goods smuggled from Colombia. The value of exports has been about one-fourth that of imports so that Venezuela is a net importer of agricultural products. But even in 1973 the deficit was only Bs628 million.

Geography of Production

The physical geography of the country, and centuries of occupation, define the agricultural regions of Venezuela. There are five regions. One is the Maracaibo basin, rich in agriculture and petroleum. The second is the coastal mountain chain, running east-west and containing agriculture and urban centers in the mountain valleys. The Andes mountains is the third region. They run from the Colombian border to the northeast, merging with the coastal region. The Andes are agricultural. The vast llanos is the fourth region. They are south of the coastal region, south and east of the Andes, and north of the Orinoco River. In the past, and still to some degree today, the llanos is cattle country. The Guayana is south of the Orinoco, a large unknown zone of mountains, rivers and plateaus.[3]

Because of this geography agriculture has tended to be concentrated in the Andes, the Maracaibo basin, and in the coastal range. Little production of any kind exists in the delta of the Orinoco or in the Guayana. The llanos is still used for cattle grazing but there are few crops. Corn, the most widely diffused crop, is grown in all of the northern and western states, with the greatest concentration in the Andean states, Lara, Portuguesa, Yaracuy, Carabobo and Miranda.[4]

Rice grows primarily in Barinas, Portuguesa and Cojedes. Potatoes are a crop of the Andes, Yaracuy, Carabobo and Aragua. The greatest concentrations of cacao are in Miranda, Sucre, and to some extent Aragua, while coffee production is more ubiquitous through the mountain states. Sesame is a commercial crop produced primarily in Portuguesa. Yaracuy produced more than one-half of the tobacco, while Anzoátegui produced most of the cotton. The expansion of cultivated land since 1950 has occurred in existing growing areas, so that the geography of agricultural production has not changed significantly in the last quarter century.

The capital region, consisting of the metropolitan Caracas zone, is primarily urban with very little agricultural production. The central region—the four major states of Aragua, Carabobo, Cojedes and Guárico—surround the capital. Aragua and Carabobo are small states with coastlines to the west of the capital, mountainous in the north and reaching down to the llanos in the south. They share Lake Valencia, second largest in the country. The two major cities, Valencia and Maracay are major industrial cities and centers of the rich surrounding agricultural zone, accounting for more than 12 percent of national production. Cojedes and Guárico are large interior states of the llanos. Agriculture, in the form of cattle raising, cotton and tobacco, is the major activity, contributing 10 percent of total output. Much of the land is vacant or is in pastures and forests.

Four states—Portuguesa, Falcón, Lara and Yaracuy—constitute the west central region. They are agricultural states contributing more than a fifth of agricultural production, including rice, tobacco, sesame, sisal, onions, tomatoes and many other crops. Zulia, a region all by itself, is the richest agricultural area (9 percent of output) including nearly every crop and animal.

The Andean states—Trujillo, Mérida and Táchira—and the state of Barinas are poor and primarily agricultural. Farming in the Andean states is in the high mountain valleys, but erosion over the centuries has taken a tremendous toll. In the nineteenth century the Andean states were the chief center for coffee production but the decline of that crop has left the zone impoverished, making do with a variety of agricultural products. The three states contribute 12 percent of national production. Barinas is an adjoining llanos state, much of it empty or host to a few cows. In the western part of the state, next to the mountains, agricultural production has thrived since the building of a highway joining San Cristóbal in Táchira with Barquisimeto in Lara, both major cities.

The northeastern region—the states of Anzoátegui, Monagas, Sucre and Nueva Esparta—produces cotton, beans, sweet potatoes, cacao and coffee (especially in Sucre) and other crops. It is a poor region, especially the islands constituting Nueva Esparta, which produces almost nothing. Monagas, a llanos state, is also poor. The four states combined contribute 16 percent of agricultural production. The huge southern and Guayana regions, more than one-half of the national territory, have few people and almost no agricultural production.[5]

Agricultural Policy and Reform

The first sustained effort to improve agriculture began in 1959 with democratic government. The program had many dimensions: accelerated government spending in agriculture for infrastructure and credit, protection against imports, minimum agricultural wages and unionization, subsidies and minimum prices, improved transportation, storage and marketing facilities, technical assistance, fertilizer and productivity programs, education and training, housing and community building, land improvements to increase land in use, expropriation, purchase, and use of government land for resettlement of rural families. It was a massive coordinated policy to modernize agriculture.[6]

The government resettled 136,000 rural families between 1960 and 1974, providing them with land and facilities to become productive agriculturalists. In some cases the government bought the land, paying the owners a negotiated price in cash and bonds. In other cases, the government turned over its own land. The peasant received title to the property—but he could not sell it—received credit, housing, land improvements, community environment, and other facilities. While some of the peasants have left their new lands, the program has had some effect on the land tenure system.

The agrarian reform program has worked primarily in crop production, and in that part of agriculture the proportion under the program crept up year by year throughout the 1960s, reaching its peak of 24 percent in 1969. The physical expansion of the program halted at the end of the 1960s and has subsequently contracted somewhat. In 1973 only 18 percent of crop production was in the agrarian reform program, the land in use declined slightly, and only 45,000 peasant families were resettled between 1970 and 1974.

With the increased spending in agriculture, the government has built many new farmer-to-market roads, improved and irrigated land, provided storage and marketing facilities on a much greater scale, and provided a much more ample framework within which market forces can influence production. At the same time, government programs have raised wages without increases in productivity, guaranteed prices to agriculturalists, provided credit (which producers have often used for consumption rather than for production purposes) and failed to press home the use of fertilizers and modern agricultural techniques. The result has been high-cost agricultural produce for consumers without a major impact on rural poverty.

As a short-term reform program to restructure and modernize agriculture, the policy has failed. Agriculture is still much the same as it was in 1960. Many peasants abandoned the land, and in fact the government redistributed relatively little land. The basic land tenure system remains in tact and most of agriculture still remains outside the agrarian program. The complex policy machinery, involving several ministries and autonomous institutes, has failed to coordinate its activities fully, initiated and dropped programs in mid-stream, failed to monitor projects, managed poorly, and wasted a great deal of money. Changes in government priorities, in personalities heading agencies, as well as disjointed activities and lack of trained personnel stultified progress in the program.

The part of the program directed at commercial and medium-sized farms worked rather well. These producers needed capital, infrastructure, and a guaranteed market, already having the motivation and a market outlook. The program provided that which producers lacked and stimulated increased output. This part of agriculture, however, was least in need of agrarian reform. The part most in need, the peasant sector, was more intractable. The Venezuelan peasant—ignorant, unresponsive to the market, and unmotivated—displayed great dedication to the status quo. Even the modest accomplishments of the program in this sector have been an important victory.

If agricultural policy is viewed as a first step requiring decades to complete, then the trend reversal in 1960 is an historic event of great importance. The government increasingly sees the program with this long-run vision, and continues to emphasize the high priority of agriculture in spite of the difficulties. The inadequacies and slowness of the first fifteen years the government accepts as a part of the learning process in the development and implementation of a many faceted policy to effect structural change.[7]

Land

Of the total area of Venezuela (91,205,000 square kilometers) only 1,103,000 square kilometers, a little more than 1 percent, was in use for crops in 1960. The most important crop usage was coffee and cacao, 45 percent of the total. Another 30 percent was under cereals. By 1959 crop land had expanded 9 percent, with coffee and cacao declining to 43 percent and cereals to 26 percent. On the other hand, land use for fibers and oil-bearing crops had expanded to 10 percent. The 1950s witnessed modest expansion and change.

In the 1960s the change was dramatic. Total land in use for crops increased to 1,837,000 square kilometers by 1969, an expansion of 52 percent. Cereals became the predominant land user, increasing its share to 41 percent, while coffee and cacao dropped to 27 percent. In effect, these two major usages switched places in the decade, with land for fibers and oil-bearing crops continuing to increase its share modestly.

In the 1970s land in use fell. By 1973 only 1,550,000 square kilometers were in use, 84 percent of peak land use in 1969. Most of the fall is attributable to the use of land for cereals, which dropped 32 percent. Coffee and cacao used the same amount of land, and more land was used for beans, fruits and vegetables. In 1974 land in use was back up to 1,793,000 square kilometers with much of the addition in cereals and legumes.

The increase in land utilization in the 1960s is largely attributable to the more vigorous agrarian land reform program of that decade. Much of the new land in use resulted directly from that program. The reversal of the expansionist trend in 1970 reflects a dimunition of the effectiveness of the program and less vigorous implementation of agricultural policy. The period of decline in land use, be it noted, corresponds to the period when Acción Democrática, the party which has traditionally pushed agrarian reform, was not in power.

The yield of agricultural land has increased. Most of the gains in the value of production per hectare were in the 1950s when land in use did not increase, and in the 1970s when it was declining. Yield was stationary in the 1960s when land in use was increasing. The overall gain between 1950 and 1969 was 40 percent, with another 20 percent between 1969 and 1973. The earlier gain in yield was all by 1955. On the agrarian reform settlements in the 1960s, however, an upward trend in yields prevailed during the 1960s, especially in the first few years of the program. The increases in output are a result, therefore, both of more land and better utilization of the land.

For some crops, yields were higher in the 1960s than in the 1950s. Corn, a major focus of the agrarian reform program, had improving yields, and yields in the program were superior to the rest of corn production. Roots and tubers, and fruits and vegetables also experienced increasing yields. Most crops followed the general yield trends, however, and there are no examples of striking increases or decreases in yields for major crops.

Cultivated pastures use more land than crops, and pasture land has more than doubled in the past quarter century. Cattle and meat production, milk and milk products, and other animal products have expanded production more rapidly than pasture land. The livestock herd has steadily increased and by 1973 was 40 percent greater than in 1950.

Venezuela underutilizes its land and despite the agrarian reform program there has been no sustained increase in land in use or yield for major crops. Much of this is due to a relatively small amount of arable land, relatively poor soils, and centuries of abuse of the land. The agrarian reform program did halt the deterioriation but it was not on a sufficiently large scale to continue to reclaim land and raise average yields systematically. In this, as in other aspects of agrarian reform, real improvement is the work of decades.

Land Tenure

A combination of *latifundio* and *minifundio* constitute the Venezuelan land tenure system. Much of the best land is in large estates, the haciendas. Absentee ownership, low land utilization, technological backwardness, low productivity, debt peonage, and lack of responsiveness to market forces are the principal features of these haciendas. The small plots have even lower productivity, and many are so small that the peasant gleans only a miserable subsistence from them. The peasant often works a tiny plot until he exhausts the land and then moves on to another plot.

A small number of agriculturalists own their own property and farm it themselves. Hacienda owners characteristically provide only the loosest sort of supervision, but many small farmers own the land and do much of the work themselves. In recent years commercial farms have begun to spring up and they are also often operated by the owner. But much of the rural sector rents, shares, or simply uses land. A renter pays the owner partly in cash, partly in produce, while the sharecropper shares the produce with the owner. But there are many squatters, peasants who have no property rights and use the lands of others illegally. In the last quarter century no major structural change in land tenancy has occurred, although the agrarian reform program has had some success in increasing the number of small agricultural owner-operators.[8]

In 1950, of the 237,700 farm units, only 3,500 were properties in excess of 1,000 hectares. These haciendas, however, controlled 79 percent of the agricultural land area. At the other extreme, 168,000 farm units were properties of 5 hectares or less, but these small plots were less than 3 percent of the agricultural land. In 1961 there were 4,200 haciendas among 315,000 farm units and they controlled 72 percent of the land, while 213,000 small farms controlled still less than 3 percent. Resettled peasants, and land brought into use by agrarian reform suggest some improvement in the 1960s.

Some 200,000 families in 1960 were in subsistence farming alone, living close to the level necessary to sustain life, with no education and no knowledge of modern agriculture. They produced primarily for their own consumption with practically no surplus. These peasants were outside the market system and changes in market forces did not cause them to alter production or crops. It was to this group that the agrarian reform program addressed itself in its resettlement projects, rather than the small group of relatively prosperous campesinos and the large number of agricultural workers on the haciendas.

In 1950, 42 percent of the farm units were farmed by their owners and 15 percent by cash renters. The remaining 43 percent of farm units had no title to the land; they were squatters, although a few had obscure and probably invalid titles. By 1961 cash renters had declined to 9 percent, but the other non-owning category increased to 49 percent, while the relative number of owners remained the same. Owners were in control of 83 percent of the land in 1950, 84 percent in 1961, while the share controlled by cash renters declined from 3 to 2 percent. Other non-owners controlled 14 percent in both years. Thus, the distribution of land among categories of users changed little in the 1950s.

At the initiation of the agrarian reform program there were 350,000 campesino families, among whom 185,000 did not possess land. About 130,000 owned only small subsistence plots and the remaining 35,000 had sufficient land but not always with clear title, and they did not possess sufficient knowledge of resources to farm adequately. The goal of the program was to upgrade all three groups of rural people, but primarily the non-land owning group.

The first decade of agrarian reform witnessed only the beginning of change. Of the 159,000 farm families touched by the program, 40 percent dropped out by the end of the decade, drifting away to the cities or back to small plot farming. In the 1970s even the resettlement program slowed down. By 1969 only one-half of those still in the program had received titles. Most of the land in the program was previously idle land, purchased from the owner or taken from the public domain; thus little redistribution of land occurred, and the land tenure system in the 1970s is substantially the same as it was in the 1950s.[9]

Labor

Agriculture has employed its labor force inefficiently. Lacking appropriate land, adequate incentive, access to technology and capital, and innocent of training or education, the productivity of the agricultural labor force was and still is below that of any other sector of the economy. In part, the inefficiency also reflects hidden unemployment. Workers with zero or very low productivity can maintain themselves in subsistence agriculture with only occasional work or by relying on relatives, making little if any contribution to output. The same workers, occupied fully, could make a substantial contribution to output.

Agriculture employed 709,000 people in 1950, 41 percent of the labor force. In 1969 agriculture employed 706,000, 23 percent of the labor force. The same labor force in 1969, however, produced in real terms almost three times as much output. In the 1970s employment did not increase significantly, agriculture's share of the labor force declined, and agricultural output continued to go up. The net result of these trends is that output per worker tripled in the last quarter century.

From 1950 to 1955 the labor force increased to its maximum of 772,000 people, a 9 percent expansion. Output, however, increased 44 percent so that output per worker, even with rising employment, increased 32 percent. From 1955 to 1969 the labor force declined back to the 1950 level and in that period the output per worker went up 118 percent. This is a much better record of improvement than other sectors. In 1950 output per worker was 18 percent of the national average of output per worker, while in 1969 it was 28 percent of the average. Agriculture thus gained on other sectors but was still much below the average. The 1970s have not altered these trends. In 1972, 673,000 people worked in agriculture, slightly higher than the two previous years.

The marked improvement in output per worker does not imply that labor productivity went up to the same degree or indeed went up at all. Some shift in the mix of agricultural products may have increased the value of output without necessarily changing productivity. In addition, the amount of land in use has increased significantly and the amount of capital per worker more than doubled. Some of the increases in agricultural output must be attributed to increases in the amounts and productivity of these cooperating factors.

But even if all of the increased output were attributable to labor, the greater output per worker does not signify increased labor productivity. Labor productivity could indeed have fallen if the constant migration to the cities took primarily the high productivity workers, leaving an ample labor force but with lower productivity. Indeed, the migration in fact was primarily in the 15 to 45 year old age group, leaving behind the very young and the old whose productivity was probably lower than those who departed. The effect was to press into service those with low productivity, or those who worked part-time or not at all.

Those who left agriculture found jobs in the city with higher productivity than the average agricultural productivity, but not necessarily higher than their own productivity in agriculture. In addition, some of the newly arrived workers found no employment or displaced others who then had to seek work. Making no contribution to output, these unemployed ate away at the productivity gains of others. Furthermore, the migration necessitated investment in social infrastructure—housing, education, health expenditures—in order to provide even minimal living conditions for the newly arrived. The same expenditures in other sectors would have had a positive productivity effect.

Labor productivity in agriculture probably increased somewhat, but it was achieved at some cost to the rest of the economy and may represent wringing some of the hidden unemployment out of agriculture. It tended to equalize urban and rural wage rates for unskilled workers. Had it not been for the migration to the cities and the tripling of output per worker, more than 2,000,000 would be in the agricultural labor force. The resolution of that problem generated the serious social and economic conditions of the cities.

Capital

Official data suggests that Venezuela uses a great deal of capital stock in agriculture in relation to output, about twice as much capital per unit output, but this is misleading. Capital in agriculture includes the livestock herd. In 1950 the herd was 38 percent of capital, and the proportion increased slightly to 40 percent by 1969. Much of the remaining capital is in the form of buildings and permanent facilities, not transportation and other farm equipment. In terms of capital characteristic of modern agriculture, Venezuelan agriculture is capital deficient.

Capital in agriculture in 1950 was less than 12 percent (excluding the livestock herd) of the national capital stock, and the ratio declined throughout the 1950s. In 1959 agriculture had only 8 percent of the nation's capital. The reason was systematic underinvestment in agriculture all through the decade. The low government priority, limited public credit, and a market narrowed by import competition did not stimulate investment. While the economy as a whole invested 25 percent or more of gross territorial product, agriculture was investing 8 to 10 percent. While capital per worker increased, the increase was less than that in non-farm sectors. In the 1950s the agricultural capital deficiency deteriorated and progress toward modernization was nil.

The government assigned a much higher priority to agriculture beginning in 1960, and the proportion of capital in agriculture increased slightly during the decade, ending at about 10 percent of national capital stock. Investment in agriculture kept pace with investment in the economy as a whole, between 12 and 15 percent of product, but with a slight tendency to decline in the late 1960s. In the 1970s agricultural investment continued but its share diminished somewhat. Capital per worker gained, even gained on the economy as a whole. The 1960s and 1970s in effect brought agriculture back into nearly equivalent participation in capital formation, but did little to repair the deterioration of decades of neglect. Agriculture still does not have enough capital, especially of the type needed for modern agriculture.

Agriculture remains the bastion of private property, private enterprise and private investment. The government has not invested at all directly in agriculture, although the rising expenditures of the Ministries of Agriculture and of Public Works indicate large amounts of economic and social overhead spent by the governments. The funds for private investment in agriculture, however, most often come from the public purse. The government has not always channelled the public credit in a way that yields the most effective investment activities.

Credit

Much of the credit available to agriculture comes, of course, from buyers, merchants, company stores and specialized private institutions. Most of this credit is seasonal, to tide agriculturalists over to the next crop. Commercial banks traditionally have not loaned to agriculture but banking law changes in 1975 obliges them to invest in agriculture. The savings in agriculture, meaning the savings of the owners of haciendas and commercial farms, do not find their way back into investment in agriculture. Therefore, if there is to be investment in agriculture, the government must provide the resources.

The government during the 1950s did little. The Agricultural and

Livestock Bank provided Bs195 million in public credit to agriculture in 1950, and the figure declined every year through 1957 when Bs49 million was granted. In 1958 public credit jumped to more than it had been in 1950 and stayed high throughout the 1960s. Most of this credit, however, was short-term credit and the major recipients were commercial farms, not the peasants.

In the 1960s the government endeavored to pay more attention to the small farmer and peasant. In 1960 some 100,000 farmers received some credit, much of it based on need, rather than for investment. The recovery rate was 27 percent. Through the years the government, through the Agricultural and Livestock Bank and other government agencies, became more selective, granted fewer farmers credit, but the recovery rate improved. In 1969, for example, 29,000 peasants received credit and the government recovered 62 percent.

In the 1970s the level of government credit reached the point where it funded one-half or more of investment activities in agriculture. About two-thirds of the credit went to small and medium-sized producers, the remainder to commercial enterprises. Credit for agriculture was by no means plentiful, but the earlier stringency and limiting effect of credit on agricultural investment was overcome.

Summary

Agriculture has been and is the weakest link in the Venezuelan economy. Adversely affected by the entry of the petroleum industry, neglected by the government, and suffering the structural constraints of colonial and early national forms, land tenancy and institutions, agriculture deterioriated through 1957, despite some good crop years. The new government initiated policies and programs which halted the deterioriation, improved the situation somewhat, but did not create a modern agricultural sector.

Agricultural output nearly kept pace with the economy as a whole, but made no breakthrough. Land in use increased, the amount of capital increased, public credit increased, and infrastructure received a large dose of capital. But despite the frenetic efforts of the government, productivity has probably improved only a small amount and agriculture remains much the same as it has been—larger but not different and not yet modern. The time span, since 1960, however, is short to change the structure of an ancient agriculture. Continued application of the policies of the 1960s and early 1970s will still require much time.

NOTES

1. Standard sources, such as the *Anuario Estadístico Agropecuario*, which has detailed data on production, prices, land in use.

2. *Primer Mensaje del Presidente Carlos Andrés Pérez*, March 1975, Oficina Central de Información, 1975, pp. 155–198.

3. Levi Marrero. *Venezuela y sus Recursos*, 1964, pp. 295–383.

4. Luís Fernando Chaves. *Geografía Agraria de Venezuela*, 1963, pp. 144–226.

5. Robert Loring Allen. *Lecciones de Economía Regional*, 1972, pp. 469–492.

6. Ramón Quijada. *Reforma Agraria en Venezuela*, Caracas: Editorial Arte, 1963, is a partisan view of agrarian reform. *Reforma Agraria Venezolana*, Caracas: Universidad Central de Venezuela, 1968, contains a group of mainly critical essays.

7. *Primer Mensaje del Presidente Carlos Andres Pérez*, 1975, pp. 159–162.

8. Ramón Losada Aldana. *Venezuela: Latifundio y Subdesarrollo*, 1969, pp. 45–76.

9. Robert J. Alexander. *Agrarian Reform in Latin America*, New York: MacMillan, 1974, pp. 31–32, 51–52, 62–67, 78–80, 92–94.

Chapter XII

Petroleum

Atop the ancient drilling rig of Zumaque-1 the Venezuelan tricolor wafted in the breeze as President Carlos Andrés Pérez took possession of the petroleum industry on January 1, 1976, sixty-one and a half years after it had begun on that site. In that time Venezuela had traversed from poor agriculture to petroleum prosperity, transforming and modernizing itself in the process, all thanks to the blessing and the curse of petroleum. The good years were when the petroleum market eagerly sought Venezuelan petroleum; the bad years were when the market stagnated and the companies by-passed Venezuela. The economy moved up and down with petroleum. And now, despite nationalization and a greater measure of control over the market, Venezuela still depends on petroleum for its resources and for the dynamism of the economy.

From the second world war through the Korean war, and up to 1958, petroleum fared well, with rising export volumes and prices. The industry was in private foreign hands, which profited handsomely and expanded their facilities. The government taxed petroleum heavily and government spending stimulated the economy. In 1958 the market turned. Prices fell sharply at first and then through the 1960s declined gently. Export volume stagnated. Even with high and rising government expenditures through greater petroleum taxation, the economy slowed down. Export volume began to recover, but peaked in 1970 when the end of the industry—10 more years of reserves—was in sight. But at that critical moment, prices began to move up so that the push of petroleum continued. Than came that spectacular revolution in the petroleum market and the price increases of 1973–1975, which set the stage for the new organization and development of the petroleum industry in Venezuela.[1]

Petroleum Development Before 1950

Petroleum, known to Indians and Spaniards and thoughout Venezuelan history, had no practical value until the Diesel and internal combustion engines came along in the late nineteenth century. In Europe and the United States companies began forming to satisfy the growing demand for liquid energy. The appetite of the industrial countries was voracious, as was the oil companies' thirst for profits. The hunt for oil began in the early twentieth century and Venezuela was high on the list because of its known deposits and the dictators' eagerness for profits.

The Spanish crown applied its laws to its colonies, including the law making all mines and subsurface materials a part of the king's patrimony. With independence, the government transferred this right to the nation and successive nineteenth century governments confirmed national ownership of the subsurface. On the arrival of the oil companies, they received temporary and specific concessions, granting them the right to extract and export the nation's petroleum. The evolution of political processes in the nineteenth century vested the national chief executive with the sole authority for granting concessions.[2]

Edwin Lieuwin eloquently tells the story of the early days of petroleum. "The law was flaunted. Fraud and deception, chicanery and double dealing, were the rules of the game. Companies most adept at adjusting themselves to the ways of the corrupt government fared best; as accomplices in the crime they got the lion's share of the good oil lands and reaped the rewards of the future."[3] It was a dark page in international business history. Royalties were modest and taxes light. Late arriving United States companies fought by foul means and fair to get the big concessions, and the British and Dutch, who arrived first, grudgingly gave way. Among them they parcelled out the land and Zumaque-1 came in, in April 1914, followed by other strikes. A trickle of oil came forth.

The first exports were in 1917, of only 121,000 barrels. By 1921 crude oil production had jumped to 1,449,000 barrels. In 1922 Barrosa-2 blew in, one of the largest gushers ever, attracting worldwide attention. A new wave of concessions resulted and soon the trickle became a flood. Output was 19,900,000 barrels in 1925, and 135,200,000 in 1930, leaving Mexico and the Soviet Union behind, and surpassed only by the United States. Output dipped slightly in the Great Depression but by 1934 production was back to the 1930 level, and by 1940 production was 183,800,000 barrels. Despite the wartime slowdown in 1942 and 1943, by war's end production almost doubled and by 1950 it doubled again.

Government-oil company relations evolved slowly with the balance of power, although remaining in company hands, tipping toward the government. In the early years the companies held undisputed sway. They wrote petroleum legislation, made and unmade ministers, added strength to the dictator, and contributed to the corruption of the political system. Beginning in 1936 the government began to assert itself, seeking more revenues and greater influence, culminating in the renegotiation of concessions and the income tax of 1943. The Acción Democrática government of 1945–1948 demanded even more and introduced the 50–50 split. The companies gave ground reluctantly, fighting each move, taking full advantage of their awesome power. In 1950 they were still in charge.[4]

The impact on society and on the economy was profound. Petroleum yanked the country out of its agricultural past, and it became an oil republic. Petroleum led the economy and—through wages, local purchases and taxes, modest though they were until the 1940s—animated domestic production. Modernization began, industry made a small start, but agriculture languished as people and the focus of the economy moved to the oil fields and the cities. Trade and fiscal dependence was complete. Foreigners made the decisions which determined economic conditions.[5]

Venezuelan Petroleum in the World Setting

In 1950 Venezuela was a key element in the world petroleum industry, even though it provided only 14 percent of crude production and possessed only 9 percent of reserves. The United States dominated petroleum, with one-half of production and one-fourth of reserves. The Soviet Union had only one-half the production and two-thirds the reserves of Venezuela. The Middle East was just becoming a factor in world petroleum, and Africa was still of no consequence. In the 1940s the companies had discovered immense reserves in the Middle East, and by the 1950 these reserves were nearly double those of the United States and one-half of the world total. Production, however, for all the Middle Eastern countries was only slightly in excess of that of Venezuela, which was second in world production and first in exports.

The world's thirst for petroleum has grown tremendously in the last quarter century. World production and consumption has grown more than five-fold between 1950 and 1975 and continues to increase despite conservation measures. The demand has grown most rapidly in the high income industrial countries, especially the United States, Western Europe and Japan. United States production increased 63 percent between 1950 and 1974 and its reserves increased 36 percent, compared to an increase in world production of 536 percent and of reserves 757 percent. The United States has increasingly relied on imports, importing in 1974, 28 percent of its crude oil. Western Europe and Japan have also greatly increased imports. Petroleum has become the major item in world trade.

Many factors explain the exploding demand for petroleum. Petroleum companies kept the price deliberately low in order to encourage the growth of demand, which led to rapid growth and high profits for the companies. Partly because of low prices, as well as high incomes, people in the industrial countries have become addicted to the automobile, using cars on a vast and wasteful scale. Truck transport for cargo became a prime method of goods movement and the airplane a major form of transportation. Petroleum has become a major energy source for space heating, and industry has converted to petroleum.

Petroleum fueled the economic and industrial growth of the industrialized countries. Western Europe and Japan especially, in rebuilding their industries after the second world war converted to petroleum, and new facilities used oil, partly in response to the low price. The petrochemical and fetilizer industries blossomed and absorbed large amounts of oil. The high state of tensions between the United States and the Soviet Union has spurred the need for military readiness, not only between the contending great powers, but also throughout the world, resulting in further heavy demands for oil.

While consumption has become centered in the industrial countries, the geography of production has shifted. The Middle East and Africa produced only 20 percent of world oil in 1950 but in 1974 their contribution was 49 percent. Similarly, world reserves have become concentrated in the Middle East and Africa. These two areas had 72 percent of reserves in 1972, but only 51 percent in 1950. Trade, therefore, increasingly flowed from the underdeveloped countries of the Middle East and Africa to the industrial countries of Western Europe, to the United States and to Japan.

Venezuela's position has deteriorated in the world petroleum market. Crude production increased only one-fourth as fast as world production. Production reached its peak in 1970 when it was only 8 percent of world production, and in 1974 when it was only 5 percent. Four countries produce more than Venezuela: the United States, Saudi Arabia, Iran and the Soviet Union. Venezuela's reserves have been stationary, with a tendency to decline. Its share in 1960 was seven percent, and in 1974 Venezuela had less than three percent of world reserves, less than the reserves of ten countries.

Venezuela failed to make the same progress as other producing areas because of its high costs relative to those of other producers, because of conditions the oil companies regarded as inhospitable, because of more limited reserves, and more recently because of deliberate government policy to limit production. Well-head costs in the Middle East and Africa are substantially below Venezuelan costs. In addition, falling transport costs because of supertankers have favored Middle Eastern and African producers. The oil companies, most of which operated on the worldwide basis, tended to shift production to the area where profit and growth prospects were the greatest. Since Venezuela was the first to apply great fiscal pressure and adopt a strongly nationalistic stance, it lost favor with the oil companies.[6]

Petroleum Operations

Peak Venezuelan crude production in 1970 was 2.5 times production in 1950, and in 1974 production was back down to the level of the late 1950s and early 1960s. The oil boom years were 1950–1957 when production increased 85 percent in a nearly uniterrupted upward movement. Production fell in 1958. The recovery in 1959 was followed by small increases each year through 1965, for a total increase of 25 percent. A slight dip occurred in 1966, but the years from 1967 to 1970 registered increases which totalled 5 percent. From top production of 3.7 million barrels a day in 1970, production fell to 3.0 million barrels in 1974, with more cuts scheduled for 1975 and later.

Before the nationalization, Venezuelan producing companies were domestic corporations owned by large international oil companies. The largest company was Creole Petroleum, a subsidiary of Exxon, formerly Standard Oil Company, a North American enterprise. This company alone has accounted for about 40 percent of crude production in recent years. Shell Petroleum, a subsidiary of British and Dutch-owned Royal Dutch Shell, is also a large enterprise with production about one-fourth of the total. Mene Grande, owned by Gulf, Exxon and Royal Dutch Shell, is third with less than one-half the production of Shell. Fifteen foreign firms and one small Venezuelan firm produce small amounts.

The area under concession to private petroleum companies contracted by nearly two-thirds between 1950 and 1975. In 1950 some 6.4 million hectares were under private concession. The oil companies, however, returned land if the prospects for oil were not good, since the government taxed the concession. Between 1950 and 1955 concession lands declined to 5.9 million hectares. In 1956 and 1957 the government granted more concessions so that by the end of 1957 the amount of concession land jumped to 6.7 million hectares. The government granted no further private concessions and the lands so dedicated gradually dwindled to 2.2 million, and continued to contract until nationalization. Beginning in 1960, however, the government did grant concessions to its public enterprise, the Venezuelan Petroleum Company, and in 1972 its concessions were 1.2 million hectares.[7]

The number of wells drilled (another important indicator of petroleum activity,) also declined. In the 1950s, 11,089 wells were drilled, including 1,277 exploratory wells. In the 1960s, however, the companies drilled only 4,917 wells, of which only 519 were exploratory wells. The 1970s witnessed a continuation of the low drilling activities of the 1960s. The success rate has remained about the same, 85 to 90 percent, but the amount of production per well has gradually increased, by more than one-half in the last twenty-five years.

The contribution of the petroleum sector to gross territorial product has been substantial but declining in real terms since 1957. In the constant prices of 1957, the highest proportion, 32 percent, was recorded in 1952. Since then gross territorial product in constant prices has grown more rapidly than has the petroleum sector. In 1969 the proportion was down to 21 percent. The same trend continued in the 1970s. Based on 1968 prices petroleum was 15 percent of gross territorial prices in 1972, and 13 percent in 1974.

The real growth of petroleum compared to gross territorial product, however, understates the continuing critical role of petroleum in the national economy. Reducing gross territorial product to real terms with constant prices omits the special importance of the externally determined petroleum prices. Imports, for example, are vital to the operation of the domestic economy. The value of exports, that is, price times quantity, determines, in part, the level of imports and hence the level of operations of the economy. In addition, petroleum prices determine royalties and taxes, and therefore public expenditures. The domestic economy feeds off these public expenditures and when they go up, so does the economy.

The petroleum sector has increasing direct ties to the national economy through sales to and purchases from local enterprise. An increasing proportion of sales of refined products is for national industry, agriculture and transportation. The petrochemical industry depends on the petroleum industry. Petroleum prices are an important determinant of investment in the petroleum sector and indirectly in the rest of the economy. Hence, despite the apparent declining petroleum sector, it remains the dominant sector of the economy.

Indeed, a comparison in current prices demonstrates its increasing importance. With the falling prices of the 1960s, the current price comparison for those years shows a decline in the petroleum sector relative to the economy as a whole. In the 1970s, however, with rising prices, a different picture emerges. While the 1957 contribution of 30 percent declined to 18 percent in 1970, by 1973 petroleum in current prices was 22 percent of gross territorial product, and 37 percent in 1974.

Refined petroleum products have followed a different trajectory from that of crude. In 1950 refined products were only 17 percent of crude, but that proportion increased each year until 1966 when it was 35 percent. The ratio stayed at about one-third until the early 1970s when it began to increase again, and in 1974 refined products were 40 percent of crude production. The peak year for refined products was 1973, a figure nearly equalled in 1970 and 1971. The peak year for crude and refined products together was 1970. In the last quarter century natural gas production has more than tripled and the ratio of utilized to total gas has increased strongly, reaching 79 percent in 1974.

Refinery capacity has also increased rapidly. Despite the fact that 30 percent of Venezuelan crude goes to the large Creole and Shell refineries in Curacao and Aruba, Venezuelan domestic capacity rose from 254,000 barrels per day in 1950, to 1,554,000 barrels per day in 1972. Additions to capacity slowed down in the 1960s, but in 1972 a 14 percent addition to capacity came on line. The two largest refinery complexes, 74 percent of total domestic capacity, are those of Creole and Shell. Ven-Gulf and Mobil each have installations of more than 100,000 barrels per day, and Chevron, Sinclair, Texas and Phillips have small refineries. The state enterprise, the Venezuelan Petroleum Corporation, has modest refinery facilities.[8]

Venezuelan crude oil is primarily heavy oil and has a high sulphur content. Refined products tend to be the heavy fractions. Fuel oil has been 60 percent or more of total refined products, with the remainder scattered over the entire range of products. With the increasing concern over pollution in the 1960s, the high sulphur content became a problem and the two major companies built desulphurication facilities in the late 1960s in order to retain the East Coast U. S. market.[9]

The foreign petroleum sector has prospered financially in the last quarter century despite the increasing fiscal pressure of the government. Gross income of the petroleum companies was Bs2,394 million in 1947, and profits after taxes was Bs745 million, which represented 48 percent of net income. In that year profits were 31 percent of income, 21 percent of average capital, and 23 percent of net fixed assets. Between 1947 and 1973 gross income went up 8 times, profits after taxes 3.8 times, and taxes

15.2 times. In 1974 alone gross income jumped by 228 percent, but profits declined 15 percent.

The split between the companies and the government has changed radically. In the 1950s, before 1958, the government received slightly over one-half of petroleum gross income, reflecting the "fifty-fifty" policy of 1948. Since 1958 the share going to the government has risen markedly. In 1958 alone there was a discontinuous jump to 65 percent for the government, the next year it was 68 percent, followed by nine years in which the government share was between these two figures. In 1969 it moved up to 71 percent, in 1971 to 78 percent, and then the highest government participation in 1972, 87 percent. In 1973 it fell back to 82 percent.

The high taxes, however, did not prevent the companies from earning ample profits. The average ratio of profits to gross income from 1958 to 1973, the period of maximum tax pressure on the companies, was 19 percent. The profit ratio varied from 9 percent in 1972 to 25 percent in 1965. Before 1958 the profit-gross income ratio was, of course, even higher than after that date. The ratio of profits to average capital invested was also handsome, ranging from 12 percent in 1960 to 38 percent in 1973. Total profits from 1958 through 1973 were 2.5 times the net accumulated investment in 1973.

Exports and Prices

The volume of petroleum production and exports increased from 1950 to 1970 and has declined since then. Crude production for export topped at 3.708 million barrels a day in 1970, 2.5 times production in 1950. From 1950 to 1957 the trend was up, and expansion in those years of 86 percent. Then production stagnated, increasing slightly in 1961, and an increase of 10 percent in 1962. The decade of the 1960s was a period of slowly rising production, but with ups and downs, and acceleration toward the end of the decade. Exports of refined products tended to increase greatly from 1950 through 1965, some increase for each year for an overall quadrupling of exports. Since 1965 exports have not expanded so much. The general pattern of exports was rapid growth in the 1950s to 1957, followed by slow growth to 1970, and decline since then.

The value of petroleum exports repeats the same pattern except that the behavior of prices enhanced the boom of the 1950s, slowed even more the slow growth of the 1960s, and despite declining volume pushed values very high in the 1970s. In 1950 Venezuela exported Bs3,357 million. In

1974 the figure was Bs60,334 million. The increase was large and sustained from 1950 to 1957, but from 1958 through 1963 export values were below the 1957 level. In 1964 there was a discontinuity in bolívar export values because of the devaluation, but dollar export values that year were almost unchanged. From 1964 to 1969 export values stagnated, hovering around Bs10,000 million. In the 1970s export values increased rapidly, with the greatest increases in 1973 and 1974, quadrupling between 1972 and 1974 when volume declined by 12 percent.

The two largest customers of Venezuelan crude oil have been the United States and Canada. In 1950 the United States alone bought 23 percent and Canada another 7 percent. These figures, however, understate the magnitude of the purchases of these two countries since they also bought the crude which Venezuela ships to Aruba and Curacao where Exxon and Shell have large refineries. In 1950 these two facilities absorbed 61 percent of all exported Venezuelan crude. The refined products are then sent to the United States and Canada. France, Argentina and Holland were important crude customers in 1950.

By 1960 the picture had begun to change. Venezuela exported only 35 percent to Aruba and Curacao, a reflection of Venzuela's own increasing refinery capacity. The United States share of direct crude exports was almost unchanged but the Canadian share increased to 10 percent. Puerto Rico and the English Antilles had become major crude purchasers, as had Brazil and Argentina. British imports of Venezuelan crude increased sharply, to 7 percent in 1960, and France was joined by West Germany as important buyers.[10]

The 1960s witnessed the stagnation of the United States market for Venezuelan crude, a result of a more sluggish petroleum market, U.S. import restrictions, and increasing Middle Eastern competition. The amounts imported remained the same but the percentage declined, so that in 1970 Venezuela shipped only 17 percent of its crude to the United States. The same stagnation happened in shipments to Aruba and Curacao which took 32 percent. The gains in the Canadian market made up in part for the lost markets, since its share increased to 14 percent in 1970. In the 1970s the market shares remained about the same.

The United States has also been Venezuela's principal buyer of refined products. In 1950 the United States share was 37 percent, increasing to 53 percent in 1960, and to 69 percent in 1970. Some of this increase is more apparent than real, reflecting increasing direct shipments of products from Venezuela rather than indirectly through Aruba and Curacao. Canada's share also increased from 4 to 7 percent between 1960 and 1970. Puerto Rico, with negligible imports in 1950, emerged as a major buyer in 1970, as did Panama and the Canal Zone. Holland, Great Britain, Spain and other European countries which had been major buyers in the 1960s were not important in the 1970s.

Since much of the Aruba and Curacao refinery output goes to the United States—55 percent in 1973—that country has clearly been the most important buyer of Venezuelan petroleum, crude and products. The stagnation in shipments in the 1960s forced some diversification of markets, with more sales to more countries, especially in Europe and Latin America. In addition, increasing sales to Canada offset lost sales in the United States market, sales lost in some cases to Canada. The rapidly increasing fuel oil sales for East Coast heating, however, confirmed the dependence of Venezuela on the United States market.

One of the large and growing uses of Venezuelan petroleum has been for internal consumption. In 1950 only 2 percent of production went to the domestic market but this grew to 5 percent in 1972, a growth equivalent to the growth of export markets. The domestic market for petroleum products, nearly one-half of which is for gasoline, is for domestic transportation and industry, to maritime transport, and to the petroleum industry itself.

Because crude oil and refined products come in many different weights and grades, and because prices reflect different purposes, there are a number of measures of price. One of the most important is realized prices, that which petroleum actually fetches in the market. An index of realized prices is the unit value of exports, that is, export value divided by volume. By this index the price of petroleum has gone through a complete cycle in the last quarter century.[11]

The unit value of petroleum exports rose during the 1950s and reached a peak of Bs52 per cubic meter in 1957. In 1958 the price broke and went down by almost 6 percent. The next year it declined further, another 11 percent, and in 1960 yet another 5 percent. Through the rest of the 1960s the unit value of exports stayed low and edged downward slowly. By 1963 the realized price index was 76 percent of that of 1957. Measured in bolívares, unit export value jumped in 1964 because of devaluation, but there was a slight decline when measured in dollars. From 1964 to 1970 export unit values continued to skid, and over the 1958–1970 period they dropped by more than one-fourth.

Realized prices began to climb in the 1970s. In 1971 alone, unit export value increased 27 percent, placing prices somewhat above the 1957 level for the first time. In 1973 unit export value increased 43 percent, and in 1974, 318 percent, a two-year increase of 4.5 times. The pattern is thus a significant increase, followed by a sharp decline in the late 1950s and then a more gentle decline for nearly a decade. Prices went up gradually at first in the 1970s, followed by the dramatic increases of 1973 and 1974.

Realized prices, based on the balance sheets of the companies, reveal a similar trend. Measured in dollars, this price measure increased from $2.11 per barrel for petroleum products in 1950, to $2.65 in 1957. From that point on prices declined until 1969 when they hit $1.81 per barrel, a decline for 32 percent. The upward trend started in 1970. Posted prices—the price oil companies quote—remained the same in the 1950s to 1957 when it advanced. Then it fell back to the 1957 posted price where it remained the same, for crude, until 1974. Reference prices from 1966 to 1970, and tax export values since 1970, are the prices on which the government collects revenues. They have been in advance of realized prices but have followed the same pattern.[12]

Investment, Capital and Employment

The value of the stock of capital in the petroleum industry increased by 106 percent between 1950 and the beginning of 1970, measured in 1957 prices. Two trends were clear in this period. From 1950 to 1960 capital stock rose by 169 percent, but from 1960 through 1969 it fell by 24 percent. In the early 1970s capital stock began rising once again, but modestly. Total assets of the companies increased year by year until 1959 when their value was Bs13.5 billion in current prices. The following eight years assets declined, but in 1968 began to rise again, not reaching the 1959 figure, however, until 1973.

Through 1972 the value of gross accumulated investment in the petroleum industry was Bs24.7 billion, a sum which was almost triple the amount in 1950. The accumulation was rapid in the 1950s and had reached Bs19.0 billion by 1960. Net investment was high in the 1950s, a total of Bs11.1 billion. Net investment declined in the 1960s; in the decade net investment was Bs7.9 billion, of which Bs2.8 million was in 1968 and 1969. From 1960 through 1967 depreciation exceeded net investment each year. The new investments in 1968 and 1969 were the desulphurication plants. After 1969 new investment continued at a declining rate, barely in excess of depreciation.

New investment dropped off sharply after 1958, the peak year in which the 1956–1957 concessions were still under exploration and exploitation. By 1962, the lowest year, net investment was only 27 percent of the 1958 level. From 1963 on, new investment increased somewhat, but with setbacks in 1966 and 1967. In 1969, the next high, it still was only 88 percent of 1958 new investment.

The disinvestment of the 1960s accompanied a shift in the nature of the capital stock. In 1960 only 42 percent of the petroleum sector's facilities were in permanent construction and improvements, while the rest was in machinery and equipment. By the end of the decade new machinery and equipment and replacements had not kept pace with permanent construction, which constituted 77 percent of the stock of capital. Nearly two-thirds of the capital stock was in refinery facilities.

The petroleum sector has had a very small labor force. In 1950 employment was 45,000 and it remained at that level for four years. Then it began to decline, which has continued every year since, except for 1956–1958 when, because of new concessions, it went up to the level of the early 1950s again. By 1970 employment in the industry was down to 24,000, less than one percent of the labor force. The number of foreigners declined sharply in the 1960s. Before 1960 more than 5,000 foreigners were regularly employed. By 1971 their numbers had dropped to less than 1,000, and by 1975 only 500 foreigners worked in the industry.

With output growing slowly in the 1960s, employment falling off, and disinvestment, the capital-output ratio declined and the output per worker increased, as did the amount of capital per worker. Output per worker doubled in the 1950s and doubled again in the 1960s. The wage bill increased by one-half between 1950 and 1969, but average pay increased 2.7 times. Operating and capital costs, as a percentage of gross income, fell as taxes increased and the companies attempted, with considerable success, to protect profits. In the 1950s non-tax costs were 40 percent or more of income, but in the 1960s this ratio fell to 30 percent or less.

Capital formation and employment in the petroleum sector reflect the pressures and opportunities facing the companies in recent years. Employment fell as exploration, the main user of labor, slowed down. New investment declined with mounting fiscal pressure and ample supplies elsewhere in the 1960s, as well as Venezuela's no-more-concessions policy. When the sluggish market began to revive in the early 1970s, capital formation and new investment did not move again, however, since the intentions of Venezuelan policy were clear. Tax pressure was great, profits and growth prospects were more appetizing elsewhere, and Venezuelan nationalization was in the wind. Despite the large gross accumulated investment, about $6 billion at the current exchange rate, the outstanding net investment in 1974 was about $1.7 billion.[13]

Venezuelan Petroleum Corporation

Established in 1960 as a part of the government's effort to capture more control over its petroleum, the Venezuelan Petroleum Corporation was the intended inheritor of the petroleum industry. Between 1960 and

1983, when the concessions were to begin expiring, the company was to be the public producing, refining and marketing enterprise, charged with increasing responsibilities in all aspects of the operation of the petroleum sector, but with no supervision or control over the existing oil companies.

The government has felt that it needed to be in the petroleum business, if for no other reason than to indicate that it had the right. But there were other reasons. It also felt that it needed a performance standard under its own control with which to measure and judge the performance of the private oil companies. The government also needed a training ground for labor, technicians and executives, people who would eventually be charged with operating the nationalized industry. The government belived that it would be desirable to have a pre-existing organization framework to take over the private industry. A part of the public company's job was to negotiate and manage service contracts, the method chosen in the 1960s to continue to cooperate with the private sector without concessions.

The company has operated on a small scale, expanding each year and fulfilling some but not all of its intended purposes. While it holds more than one-third of all concession land, it contributed only slightly more than one percent of crude production and has not fully explored its concessions. It has less than two percent of refining capacity and produces an even smaller proportion of production. One of its principal activities has been in the domestic marketing of gasoline and other petroleum products. At the end of 1972 more than one-third of the gasoline sold in the country came from its retail stations and eventually it will have a domestic monopoly. It also transports and markets natural gas.

In the early 1960s the government, having decided that concessions were no longer acceptable, proposed to substitute service contracts and charged the company with formulating, negotiating and supervising the contracts. The exploration, exploitation and export of petroleum would be in private hands under contract. The idea was to pay the private company only the supply price of the capital and technical assistance plus a small risk premium, as specified in the contract, with the government pocketing the difference between export price and contracted costs.[14]

Service contracts were slow in coming. The legislature did not get around to authorizing them until 1967, and even then they remained largely undefined. The private companies demonstrated a notable lack of enthusiasm; they hoped more concessions would be forthcoming. In addition, the public company was inexperienced and did not know exactly what to do. It was also anxious to get its own production facilities into operation first. By 1971 only five service contracts had been negotiated and no major strikes had resulted.

When nationalization was under serious consideration in 1974 and 1975 the government decided that the existing public company had been plagued by problems, lack of leadership, and unfulfilled plans. The government decided to start afresh, not with another huge operating company, but rather wtih a holding company, Petróleos de Venezuela (Petroven for short) which would own all the operating companies and make industry-wide decisions. Each of the existing private companies would be a unit in the new arrangement and would be taken over intact, but supplanting the private executives with public officials. Even the public company would be a part of Petroven.

Under the provisions of the nationalization law Petroven can contract and form associations with the private companies. The need for such assistance is greatest in marketing abroad, in the acquisition of technology, and in technical assistance. In early 1976 Petroven had already contracted with the private companies for the marketing of Venezuelan petroleum, and negotiations were under way for a continued stream of technology to Venezuelan operating units.[15]

Organization of Petroleum Exporting Countries

In Bagdad in September 1960, four Middle Eastern countries—Iran, Iraq, Saudi Arabia and Kuwait—and Venezuela established the Organization of Petroleum Exporting Countries (OPEC). Other countries subsequently joined: Qatar in 1961, Libya and Indonesia in 1962, Abu Dhabi in 1967, Algeria, Nigeria, Trinidad-Tobago in 1972, and Ecuador in 1973. OPEC is an intergovernmental organization of the major crude petroleum exporting countries, with permanent headquarters in Vienna.

Venezuela was one of the prime movers in the initiation of OPEC, and a strong supporter throughout its development. The idea of producer cooperation is, of course, very old but it has seldom moved beyond the talking stage, much less to the spectacular success of OPEC. Dr. Juan Pablo Pérez Alfonzo, the minister of Hydrocarbons and Mines in the new constitutional government, was a key figure and his outlook was influential in the formative years of OPEC. Venezuela's more advanced nationalistic oil policies led the way and set the tone of OPEC during most of the 1960s. More recently the relative weight and more aggressive stance of some of the Arab countries and Iran has begun to be felt and Venezuela's influence in the organization, while still important, has waned.

The principal function of OPEC, according to the Bagdad Agreement, "will be the unification of the petroleum policies of member countries and the determination of the best means of safeguarding the interests of member countries individually and collectively."[16] Member countries regarded OPEC as a counterweight to the influence exercised by the giant North American and European oil companies whose practice it was to deal with each producing country separately, bringing their power to bear on each one in turn, thereby extracting the best treatment from each. OPEC was to redress this power imbalance.

The 1960s were the learning years. Despite the fact that they were major oil producers, the member countries did not know the petroleum business. The oil ministers and their aides met and exchanged ideas, taught one another, and learned how things were done in one another's countries. OPEC formed a technical staff. Venezuela's experience was invaluable at this stage. Policy coordination was initiated, often on the basis of existing Venezuelan policy, such as the "fifty-fifty" split, conservation, reference prices, and techniques of negotiations with the companies.

No question existed about the latent power of OPEC. Its members in 1971 produced 53 percent of world crude, exported 83 percent of net crude exports, possessed 72 percent of world petroleum reserves. The question was whether or not they could act collectively and thereby influence the oil companies. At first, the companies ignored OPEC, hoping it would go away. Later they found themselves dealing with individual countries on the basis of OPEC resolutions, each country having accepted a common position. Still later they found themselves dealing with blocks of OPEC members or all of OPEC.

The focal point of OPEC has been the price of petroleum. This amounts to the acceptance of the Venezuelan position that petroleum is an asset whose yield should be maximized rather than a product whose output and growth should be maximized. The value of production, hence price, determined government revenues. By 1970 OPEC had developed the policy instruments, the reference price, export values, sell-back price, and other means by which the companies were obliged to set prices in part in accordance with the supplier country's wishes. OPEC will, technical competence, and ability to count on the cooperation of all members reached the turning point in the early 1970s.

OPEC transformed itself into an intergovernmental producer cartel. During the 1973 Middle Eastern war, negotiations with the companies over the price of petroleum broke down. In this crisis setting, OPEC began setting the price of petroleum, and has maintained the market price close to its price by regulating supply. The control over supply, however, is incomplete, since there is no firm agreement or quota system for the members. OPEC expects each to act so that the established price remains the market price. In 1974 and 1975 member countries indeed behaved in this manner, reducing supplies in order to hold the price up.

Unabated demand and policy uncertainties among petroleum consumers helped to maintain the effectiveness of OPEC's position.

Over a short period a cartel can be successful, but over a longer period uncertainty and eventual breakdown has awaited all efforts of this kind. To be effective, supply must be controlled and demand must be little influenced by price. Producer solidarity is difficult to maintain since a given member can sometimes improve his position by cheating on supply controls. The demand side may react by seeking substitutes or finding additonal non-OPEC supplies. It is probable that the cartel will weaken over a period of time but it is certain that OPEC will continue to exercise a strong influence on price and other aspects of the world petroleum industry indefinitely.

Even if the cartel weakens over time, Venezuela justifies its own participation on the revenues it has already earned and will earn while the cartel remains in force. The grasp of the petroleum companies has been torn away from the helm of the industry, and all producers, including Venezuela, are better off and will continue to be better off than they would have been had pre-1973 conditions continued to prevail. The wealth transfer has created conditions in which development prospects are brighter than at any previous period in Venzuelan economic history.

Nationalization

The cry for nationalization began with the political liberalization in 1936 but was always held in check by financial and technical realities and by the power of the companies. Each successive government came to terms with the oil companies and satisfied itself with a somewhat larger slice of the pie and more influence in decision-making. In the constitutional regimes since 1958 the position of the government has been that petroleum is already a national industry, temporarily in the hands of foreigners. The cost of compensated nationalization would have been high, and uncompensated nationalization would involve grave political and economic risks. Furthermore, the government did not consider the country technically capable of operating and developing the industry. The industry, however, would revert, cost free, beginning in 1983, and by that time the country would be ready. Only fringe parties of the left carried on the call for nationalization.[17]

In 1970 the situation began to change, culminating in the decision to nationalize the industry in 1974. At the OPEC meeting in Venezuela in 1970 the pieces of the cartel finally came together and the prospects for petroleum price increases became better. Not by accident, at exactly the same time, the Venezuelan congress debated and passed a new income tax law in which export values (to be used in calculating taxable income) were to be determined unilaterally by the government. In 1971 and 1972 the government nationalized the natural gas industry and established the conditions of reversion, including a substantial escrow fund. The thinking was still of reversion in 1983.

Events then began to move rapidly. Under OPEC influence the price of petroleum began to go up and Treasury vaults began to fill. International reserves piled up. Reversion before 1983 began to be discussed. In the electoral campaign, which resulted in the election of Preident Carlos Andrés Pérez, early reversion or even nationalization became campaign issues. As 1974 wore on it became obvious that financial resources for nationalization would be no problem. Responsible officials advised the government that the country had the technical and organizational talent and capability to operate the industry. In December 1974, after the report of a special commission to study the petroleum industry, the government announced that it would nationalize the industry at once.

In 1975 the country debated the terms of nationalization, with near universal belief that it was the appropriate step. Congress resolved most of the issues easily. The most debated element concerned continued cooperation with the private petroleum companies. The government insisted on a provision allowing for such cooperation; COPEI and most other parties opposed it. The government prevailed. Congress passed the nationalization law in August 1975, and the President signed it. On January 1, 1976, Venezuela took possession of its petroleum industry.[18]

Mining

Mining has always been a minor activity, except for iron ore mining which began in 1950. All of mining has never made a contribution of more than 2 percent to gross territorial product. Before nationalization in 1974 the industry consisted of the work of two companies, Iron Mines Company of Venezuela—a subsidiary of Bethlehem Steel, and the Orinoco Mining Company—a subsidiary of U.S. Steel, on two concessions.[19]

Production and exports climbed rapidly in the 1950s, reaching a peak of 20.2 million tons in 1960, followed by a decade of decline and stagnation. The highest exports that year were those of Orinoco Mining, of $159 million. In the 1960s exports averaged only $103 million, an average of 4 to 5 percent of total exports. In 1970 exports exceeded the 1960 level, followed by declines for two years. Exports in 1974 were 26.4 million tons but earnings had ascended to Bs1,078 million, about $257 million. A part of the increase is due to price increases, some 25 percent in 1974 alone.

The rapid rise and subsequent levelling off at lower levels in the 1950s and 1960s is attributable to the near exhaustion of United States reserves at the end of the second world war and the subsequent search for and discovery of new reserves. The Venezuelan finds were early, and enjoyed a prosperous market before other producers came into the market. World reserves went from 54 million tons in 1950 to 200 million tons in 1967. After 1960 Venezuela lost some of the United States market to Canada and other new producers. The new-found prosperity of the 1970s was part of a general resource stringency and ever-rising demand.

Exports are not the only basis of iron ore mining in Venezuela. The deposits are in the Guayana region, easily accessible to water transport on the Orinoco River, which makes the region open to both imports and exports. Other minerals, such as bauxite, are nearby, and there is an abundance of hydroelectric power. The government steel mill was built in this region to use the iron ore, and the government and private companies established other industrial facilities in the region in the 1960s and 1970s. Ciudad Guayana has become the hub of a growing industrial complex, one of the pillars of which is iron ore.

The government has subjected the iron ore industry to the same kinds of pressure as the petroleum industry in the last quarter century. In late 1974 the government nationalized the industry, even though the concessions had twenty-five more years to run. The government negotiated with the companies and agreed on a compensation of $101 million. The existing companies will continue to play an important role under contract to the goverment.

Summary

The private petroleum industry, which made Venezuela the richest less-developed country in the world, bit the dust in 1975. Prosperity, recession, stagnation and high prosperity punctuated the preceding quarter century. When the petroleum sector prospered it led the economy and conferred prosperity on it as well. In recession, petroleum dragged the economy down with it. In stagnation, petroleum lagged the economy and slowed down growth. In prosperity again, petroleum has conveyed substantial wealth to the economy.

NOTES

1. The statistical data in this chapter were obtained primarily from the annual publication of the Office of Petroleum Economics, Ministry of Mines and Hydrocarbons, *Petróleo y Otros Datos Estadísticos*, and other standard sources.

2. A chronicle of the development of petroleum from the beginning, in 1498, to 1970, month-by-month, year-by-year, is told by Aníbal R. Martínez. *Cronología del Petróleo Venezolano*, 1970. A standard nationalistic analysis is by Romulo Betancourt. *Venezuela: Política y Petróleo*, 1956. A useful volume containing legal and legislative details is Salvador de la Plaza and others. *Breve Historia del Petróleo y su Legislación*, 1973.

3. Edwin Lieuwin. *Petroleum in Venezuela: A History*, 1954, pp. 37–38. This book is a first rate analysis of the petroleum industry to the early 1950s.

4. Rómulo Betancourt. *Venezuela: Política y Petróleo*, 1956, chapter V. Luís Vallenilla. *Auge, Declinación y Porvenir del Petróleo Venezolano*, 1973, covers this period in detail, espcially in chapters IV and V. See also the English language version, *Oil: The Making of a New Economic Order*, 1975.

5. Aníbal R. Martínez. *Nuestro Petróleo*, 1963, pp. 77–81; Eduardo A. Acosta H. *Este Petróleo Es Venezolano*, 1970, pp. 115–144.

6. Michael Tanzer. *The Political Economy of International Oil and the Underdeveloped Countries*, 1969, pp. 59–63, covers the relationship of Venezuela to Middle Eastern petroleum.

7. Pedro Esteban Mejía Alarcón. *La Industria del Petróleo en Venezuela*, 1972, pp. 117–124; César Balestrini. *La Industria Petrolera en América Latina*, 1971, pp. 200–228.

8. Cesar Balestrini. *La Industria Petrolera en América Latina*, 1971, pp. 229–233.

9. José Antonio Mayobre and Benito Raúl Losada. *La Desulfuración en Venezuela*, Caracas: Editorial Arte, 1970, pp. 21–39.

10. Hector Malavé Mata. *Petróleo y Desarrollo Económico de Venezuela*, 1962, pp. 29–43; Domingo F. Maza Zavala. *Venezuela: Una Economía Dependiente*, 1964, pp. 13–61.

11. Pedro Esteban Mejía Alarcón. *Monopolio y Precios del Petróleo*, 1964, discusses price formation; and Luís Vallenilla. *Auge, Declinación y Porvenir del Petróleo Venezolano*, 1973, brings the discussion up to date.

12. Little has been written about reference prices, which after 1970 were called export values. See Luís Vallenilla. *Auge, Declinación y Porvenir del Petróleo Venezolano*, 1973, pp. 330–332; and pp. 502–515.

13. The negotiated price will be paid partly in cash and partly in interest-bearing bonds. The international reserves of Venezuela are sufficient to have paid the entire sum in cash.

14. Juan Pablo Pérez Alfonzo. *El Pentágono Petrolero*, 1969; and by the same author, *Petróleo y Dependencia*, Caracas: Síntesis Dosmil, 1971. Dr. Pérez Alfonzo is one of the chief architects of both the Venezuelan Petroleum Corporation and the Organization of Petroleum Exporting Countries.

15. *Resumen*, 21 December 1975, pp. 10–20.

16. Cited by Luís Vallenilla. *Auge, Declinación y Porvenir del Petróleo Venezolano*, 1973, p. 279. This author has an up to date discussion of OPEC in the English-language version, *Oil: The Making of a New Economic Order*, 1975, especially in chapters VI, VII and VIV.

17. Manuel R. Egaña y Otros. *Nacionalización Petrolera en Venezuela*, Caracas: Litografía Melvin, 1971, pp. 23–40.

18. *Resumen*, 11 January 1976, pp. 3–8.

19. Henry Gómez. *La Industria del Mineral de Hierro en Venezula*, 1970; and data from the annual *Informe Económico*, Banco Central de Venezuela.

Chapter XIII

Industry

Poverty and minute markets prevented industrial development before the advent of petroleum, and the plethora of foreign exchange brought in by petroleum subsequently forestalled domestic industrial expansion in Venezuela. Coffee and cacao exports plus small-scale artisan and industrial activity were sufficient for the tiny markets for manufactured goods until the 1920s. When the market grew enough to support more industry, imports precluded domestic manufacturing. National firms, high cost because of their small size and lack of experience, and forced to pay high wages induced by petroleum, could not compete against imports.

The second world war imparted the first small impulse toward industrialization. Still, the abundant foreign exchange and cheapness of imports discouraged domestic enterprise. The market remained small—only 4.3 million low-income people—but the wartime cutoff of imports resulted in the expansion of food processing, textiles and clothing, and other light consumer goods production. Industry continued to expand modestly after the war, but at mid-century, Venezuela was still the classic case of a supplier of a primary product and buyer of industrial goods from economically more advanced countries.

Since 1950 industrial expansion has continued with industry growing faster than the economy as a whole and at an accelerating pace. Economic policy has sponsored industrialization, especially since 1958, in providing credit for expansion, by excluding imported manufactured goods, and by installing economic infrastructure. Still, high cost raw materials and capital goods (much of which had to be imported) and high wages, as well as small markets and inefficiencies bred by protection, paternalism and dated management practices has so far deferred a great industrialization drive. Industrial progress, however, has been sufficient to diversify the country's economic structure and establish the foundation for more balanced economic growth.

Industrial Growth and Composition

Industrial production, consisting of manufacturing, construction, and water and electricity, has claimed an increasing share of gross territorial product. In 1950 its contribution to total product was 17 percent in 1957 prices, with manufacturing in the lead with 10 percent, construction less than 7 percent, and electricity a minute fraction. By 1969, on the same price base, industry had edged up to 20 percent, with the manufacturing part up to 13 percent. The rapid electrification of the country raised that sector's contribution to not quite 3 percent, but construction was down to a little more than 4 percent.

In the 1970s industrial growth outpaced the economy. The industrial share was up to 23 percent by 1974 (measured in 1968 prices) and the manufacturing part was 16 percent of gross territorial product. Power and water was now consistently somewhat more than 2 percent, while construction was not quite 5 percent. While the quarter century record does not represent a great leap forward into industry, it is an impressive gradual diversification of production.[1]

The rate of growth of manufacturing during the 1950s was 12 percent per year, almost half again the rate of growth of the economy as a whole. The petroleum boom animated industry and the negligible surveillance of the dictator led to high profits, favorable expectations, and the greatest industrial expansion of the last quarter century. In the 1960s the positive efforts of the government to encourage industry were offset in part by its more militant social and economic policies, and manufacturing growth was only slightly ahead of the growth of gross territorial product. In the 1970s, excluding petroleum refining which in real terms grew slowly, manufacturing began to grow more rapidly, recording for 1970–1974 growth rates nearly double the growth of the rest of the economy.

Construction has been a highly volatile industry. In the 1950s its growth exceeded the growth of the economy slightly, but experienced wide swings in the rate of growth. In the early 1960s the recession and tight credit markets depressed the construction industry, and even the growth in the rest of the decade only brought the industry back to slightly ahead of where it was in 1959. The 1970s brought back the earlier high growth rates; in 1972, for example, construction grew 25 percent. But the old instability remained, since growth slowed to 5 percent in 1973, and in 1974 declined to only slightly more than the 1972 record. The recession and slow progress of the 1960s, however, resulted in construction growth since 1950, slightly less than the growth of the economy as a whole.

Reflecting the almost complete electrification of the country since 1950, the growth of power and water has been very high. In the 1950s the average annual increase was 19 percent; in the 1960s and 1970s, 13 percent per year. Bear in mind, however, that Venezuela started from a small base in 1950 and had the advantage of great hydroelectric potential, especially in the Guayana, and a cheap fuel, petroleum, for thermal plants. Even so, from a markedly inferior position, Venezuela now has more installed capacity and production per person that any country in Latin America.

Venezuela remains a producer of light, consumer, non-durable goods, although durable goods production has grown more rapidly. In 1950 durable goods were 23 percent of manufacturing, and in 1974 these goods accounted for 28 percent. The greatest progress was in basic metals and metal products. In non-durable goods petroleum refining was traditionally the most important industry, but in real terms in the 1970s the industry declined and food processing displaced refining as the most important industry, claiming 17 percent of all manufacturing. But even this industry had declined as a share of all manufacturing; it was 22 percent in 1950. Textiles and clothing, paper and paper products, and chemical products increased their share. Within manufacturing, diversification has occurred, with more rapid growth of new industries.

Import Substitution

The structure of tariffs in 1950 accorded sufficient protection that throughout the decade domestic industry expanded at the expense of imported industrial goods. In 1950 national industry supplied only about one-third of the total domestic market while imports provided almost one-half. Domestic agriculture supplied a little less than one-fifth of the national market. These shares shifted notably in the 1950s. In 1956 the domestic industry share was up to 57 percent and the import share was down to 26 percent, while the agriculture share remained about the same.

In 1960 the new government initiated an import substitution policy, designed not only to develop national industry but also to reduce total imports. It consisted of higher duties, quotas and other non-tariff barriers, and included a modification of the trade agreement with the United States which had inhibited protection of some industries. The result was the recapture of the ground lost to imports in the last part of the 1950s as well as further gains for domestic manufactures. Imports in 1960 were only 22 percent of the national market, industry was 59 percent, and domestic agriculture also gained slightly. The movement away from imports and to domestic industry continued during the decade and in 1969 imports were less than 11 percent of the national market, domestic industry was 69 percent, and domestic agriculture was 20 percent.[2]

The import substitution program provided a strong impulse in food processing and clothing and personal effects. In 1950 Venezuela imported 36 percent of its needs in food, beverages and tobacco; by 1969 only 8 percent came from abroad. Domestic industry supplied only 25 percent of clothing and personal effects in 1950, and the remainder came from imports. But by 1969 domestic industry provided 83 percent of the goods in this category. Before the import substitution policy Venezuela imported large quantities of cigarettes, processed meat, cheese, sanitary products, tires and tubes, paper products, and many other products. Domestic producers have replaced imports of these products.[3]

The import substitution program has continued during the 1970s. The two most important recent developments were the abrogation of the trade agreement with the United States in 1972, which unbound some tariff rates and accorded more domestic protection; and the revision of the tariff system in the same year, replacing the specific rates with *ad valorem* rates. The government deliberately established rates in the new schedule to protect infant industries and in some cases excluded imports entirely. For example, new automobiles face a 300 percent tariff, virtually prohibiting them; and auto parts for the domestic assembly plants are subject to a regime which, over time, requires a greater proportion of domestic origin.

The effectiveness of import substitution has been greatest for final consumer goods. Indeed, the policy has now almost run its course in this first and easiest step of an import substitution policy. The substitution of imported raw materials and intermediate and capital goods has barely begun. Indeed, as domestic industry has expanded, imports of inputs into national manufactures has increased. The substitution of imports of final goods did not spark a general industrialization or even move much beyond the specific protected sectors. The survival of even these domestic industries requires continued protection.[4]

In order to relieve the cost pressure of imported inputs into domestic industry, the government exonerates the duties on these imports for industries whose growth it was trying to spur. During the 1960s it collected only Bs1,035 million of Bs5,877 million owed in this exoneration program, and Bs5,372 million were exonerated from 1970 through 1974.

Had these duties been collected, the affected industries, even with protection, would have made little progress.

Dependence on imported goods remains almost as pronounced as two decades ago, although the composition has shifted from final manufactures to industrial goods for industry. Much of the visible manufactured goods is now of Venezuelan origin so that the appearance of dependence has lessened. The dependence, however, has just moved back a step in the production process, into parts and assemblies, industrial raw materials, and capital goods. The deepening of industrialization, so far, has only a slim beginning in steel and steel products, chemicals, and aluminum, and aluminum products.

Government Economic Policy

In addition to trade policies affecting industry, the government has endeavored to promote industrial expansion with other policies. The most important policies have been the construction of economic infrastructure, government purchases from industry, financial measures, including loans, stock purchases, reduced taxes and subsidies, cooperation and consultation with industry, including the installation of planning to reduce uncertainty, and in some cases the participation of the government in industrial enterprise, either on its own or jointly with the private sector. In addition, the government has refrained from pursuing some policies, such as income redistribution, socialization of industry, and hostility toward the private sector, which would have slowed industrial growth.[5]

Rising expenditures on economic infrastructure have made it possible for industrial enterprises to get a start and expand. One of the most important elements has been the building of the transportation and communications system. The government has constructed the best and most complete road network in Latin America, resulting in falling freight rates and faster and more reliable service. It has also built industrial parks with full service facilities and made them available to enterprise at a fraction of cost. Improved telephone, postal and telegraphic service has also been vital in industrial development. Without this infrastructure, industry could not have grown.

The completion of the national electric power grid, based on inexpensive hydroelectric power, has assisted domestic industry in which power is a significant input and has encouraged some industrial development, such as aluminum, which requires cheap power. In addition, the government education and technical training programs have provided an increasing flow of productive workers to domestic producers. Some programs are directed specifically at industry, but even the general educational programs raise the skill level of the labor force.

In addition to the specific government infrastructure programs to assist industry, the rising curve of government expenditures has made the government the best customer of domestic industry. A buy-Venezuelan program which enjoins ministries and agencies to purchase domestic goods when available, the connections between the bureaucracy and industry, and propinquity resulted in domestic purchases even when imports were cheaper. Recent restrictions on foreign investment reserve some lines to domestic firms and establishes such stringent guidelines that foreign firms shy away from Venezuela.

The government also helps in the financing of industrial development. Not wholly for purposes of industrial expansion, the Central Bank holds the rate of interest down, encouraging borrowing. The funds are mainly short-term money, but its cheapness encourages resort to financial institutions for working capital, thus freeing other capital for investment purposes. In addition, commercial banks buy equity capital in enterprises, yet another source of funds for investment. The proliferation of new and different kinds of financial intermediaries since 1960 has made long-term funds available directly, as well as indirectly through housing and construction. The soundness of the bolívar and maintenance of strong reserves has meant that domestic industry has faced no foreign exchange problem.

The government participates directly in financing industry through the Venezuelan Development Corporation and other programs. The Corporation, established in 1946, operated on a small scale until the 1960s when it began to expand rapidly. In 1964 its investments and loans were Bs1,289 million, and reached Bs2,200 million in 1970, and Bs2,586 million in 1974. The Industrial Bank of Venezuela, a government intermediary, grew rapidly in the 1970s; its loans tripled to Bs1,602 million between 1970 and 1974. An institute for small and medium size industry also provided financial, as well as technical assistance. Government programs did not meet all of the financial needs of industrial enterprise but were a critical element which fulfilled specific development needs.[6]

The low level of taxes, the lax manner in which the government collected taxes, and its tolerance of high profits provided ample vistas for enterprising domestic firms. The government has had no means for scrutinizing the records of firms adequately, accepts the unverified statements of firms, and seldom raises its voice against high profits, even for firms receiving government protection and assistance. Indeed, in the 1970s, 90 percent of all firms reported profits and 95 percent of firms with sales over Bs10 million reported profits.

Industrial regulation touches industry lightly. To be sure, the government has established norms and codes for working conditions, for product safety and industrial practices, but they are not rigorously enforced. The heavy-handed and omnipresent government regulation characteristic of many industrial countries is absent in Venezuela. The consumer movement has not yet reached Venezuela. While industrial enterprise is not completely free-wheeling, detailed government inhibitions are few.

Government economic planning and coordination, even though on occasion opposed by industry, benefits industry by providing guidelines to the prospective behavior of its biggest customer—the government—as well as government expectations of the private sector. It has been within the planning framework that the government has granted protection, credit, technical assistance and other industry-stimulating measures. Planning has not directed industrial growth or dictated investment, but rather has endeavored to create conditions in which the private sector would respond through self-interest to what the government considers the country's industrial development needs to be.

Direct government industrial production has been limited, and the government has never seriously entertained mounting a massive government-sponsored industrial drive. Most of industry remains the preserve of the private sector. In those instances in which the domestic private sector did not act to fill a gap the government considered important, the government has stepped in, either as the producer of its own or in partnership with the private sector. Such situations include those in which the risk was great, the capital requirements high, or a consensus existed that the activity was in fact the responsibility of government. By this process the government, over the past quarter century, has become a major industrial producer.

In the 1950s steel, petrochemicals and hydroelectric power development were under government auspices. The private sector could not muster the amount of capital required and the sentiment prevailed that large-scale projects of this type were the government's business. The government constructed the integrated steel mill to use the Guayana iron ore, and built Macagua I, a large hydroelectric facility at the juncture of the Orinoco and the Caroní. It also started a small petrochemical plant at Morón. In the 1960s the government expanded the Morón plant and built another at El Tablazo near Maracaibo, enlarged the steel mill, and built the huge Guri dam and hydroelectric plant in the Guayana. It participated in building an aluminum industry to use the cheap power.

With the nationalization of petroleum, natural gas and iron ore, government production dominates in the non-agricultural extractive industries and in the industries associated with the use of these natural resources. The government limits its appetite to these sectors and has not threatened to take over or interfere or compete with the private operation of the rest of industry, but always with the qualification that industry must serve the development needs of the country. The government has earmarked the reserve funds in the Venezuelan Investment Fund for steel and aluminum expansion, hydroelectric power, petrochemicals and shipbuilding, on the assumption that the private sector, coordinated and influenced by the government, will satisfy the country's other industrial requirements.[7]

Structure of Industry

Venezuelan industry has a distinctly modern look, since most of it came into existence in the last quarter century. It remains small-scale by the standards of industrial countries, but uses modern technology and production methods and endeavors to adhere to the product characteristics and quality of the competing imports. Many domestic firms bought entire plants abroad, make products which are almost identical to their foreign counterparts, and maintain licensing and technological relationships to foreign firms.

A pronounced trend toward increasing the size of enterprises has reduced the number of enterprises and increased their capitalization and labor force. In 1953, 16,542 industrial establishments employed an average of 13 workers each, and had average capital of Bs127,400. Through failure and consolidation the number of enterprises declined to 7,531 in 1961, of which 196 were large, 1,119 medium size, and 6,216 were small, mainly artisans employing one or a few workers. The number of employees per enterprise was 35 and each firm had Bs573,000 in capital. In the 1950s the casualties were heaviest among small enterprises, but the entire industrial structure changed radically.

The consolidation process continued during the 1960s and 1970s, although the import substitution program increased the number of firms temporarily in the early 1960s. By 1966 the total number of firms had declined, however, to 7,305 but the number of large firms increased to 325. The number of medium size firms increased but the number of small firms diminished again. In 1966 each firm had an average of 57 workers and capitalization increased to an average of Bs794,000. By 1970 only 5,945 firms existed, with more large and medium size firms and fewer small firms.

Most industries have a relatively large number of firms and are competitive, but a few industrial giants have begun to emerge. In 1970 over

780 firms were in the clothing industry, most of them with less than 50 employees. In soaps, cosmetics and medicines, as well as plastics, 330 firms had an average of 27 employees. Food processing, one of the largest industries, had 1,608 firms. Other industries, however, had few firms. There are only a few brewers, tobacco product firms, automobile firms. Over the past twenty-five years competition among the oligopolistic and monopolistically competitive firms has remained active, or even strengthened if publicity and advertising is used as a guide.

Some business interests had widespread holdings in more than one subsector of industry, and in commerce and banking as well. The Vollmer-Zuloaga interests, for example, own enterprises in cement and sugar, metallurgy and banking, beer and electric power, television and glass manufactures, among others. The activities of Eugenio Mendoza embrace paper and cement, banking and paints, and many others. A few of these conglomerates form the spinal column of the economy and these industries, amply financed, better managed, and more amply connected to the government, have prospered.

Domingo Alberto Rangel in his *La Oligarquía del Dinero* in 1972 recounts the fascinating story of some of the great business families of Venezuela and their widespread interests in every branch of economic activities. He argues that "the country belongs to twelve economic groups whose unique position and incontestable influence establish the guidelines of the country's economic and political life."[8] By his figures, which may suffer from ideological vitiation, the assets of the twelve largest industrial-commercial-financial groups were, at the end of the 1960s, more than Bs8,000 million, a little less than one-fourth of the capital stock of the country, except for petroleum, housing and government.

Most industrial enterprises are anonymous companies or anonymous societies, having attributes much like those of the corporation. Most had highly concentrated ownership. Almost none list stocks on the stock market and most had five or fewer owners. In 1973 and 1974, for example, about one-half of the new anonymous companies had one or two owners and a tiny fraction more than 50. Many enterprises are family firms, owned exclusively within the family. Some industrial enterprises continue as single proprietorships, and a form of partnership, the limited responsibility society, exists.

Caracas is the center of industry with as much as two-thirds of the country's manufacturing in the metropolitan zone. The greatest concentration of electric power production is in the Guayana region. Consumer-related small industry exists in all major population centers, but Caracas dominates in all lines of production. It has the locational advantages as the hub of the transportation network, proximity to the largest port and raw material and capital goods imports, the largest market and supply of labor. In addition, as the financial and governmental center, Caracas attracts industry. The expansion of the road network, which could have helped to decentralize industry in fact operated as a conduit through which enterprises, management and labor gravitate to Caracas.

Three secondary centers of industry have emerged in the last two decades. Valencia is a center for the automobile industry and related manufacturing. Maracaibo, the second largest city and center of petroleum operations, also expanded rapidly. The third center is Ciudad Guayana. Its focus is the steel mill, iron ore, the aluminum plant, metallurgy, and power-related industries. In the 1970s the government's program to decentralize industry has had relatively little impact.[9]

Investment and Capital Stock

The key to industrial growth is investment, and over the past twenty-five years Venezuelans have invested heavily in manufacturing and electric power. For the economy, the rate of investment during the 1950s was 25 percent, while in the 1960s it was only 14 percent, picking up again slightly in the 1970s. Even so, investment in industry in the 1960s was nearly double that of the 1950s. Average annual investment in the 1950s was a little over Bs400 million a year; in the 1960s it was about Bs700 million a year.

In the 1950s up to 1957 investment was about 10 percent of total investment. If housing and public administration are excluded from aggregate investment, the proportion would, of course, be higher. Most investment was in manufacturing. In 1957 investment in manufacturing and electric power both jumped. Investment in electric power has remained high, running one-fourth to one-half of investment in manufacturing. From 1957 to 1970 investment in industry was about 18 percent of total investment, but in 1971 the ratio increased and has subsequently been 20 percent or more.

The impact of investment on output was high in the 1950s because business was moving into easy new lines of production requiring little capital, and the manufacturing base was small. The import substitution program and the government electrification program resulted in relatively greater amounts of capital formation in industry. The import substitution impulse dwindled in the slow growth period of the late 1960s but high investment in manufacturing return in the 1970s.

Investment in industry comes overwhelmingly from Venezuelans.

What little foreign investment there was tended to be concentrated in manufacturing, with no foreign investment in electric power. The trend in the last quarter century has been toward an increasing share. In 1950 less than 2 percent of gross accumulated investment was foreign. The ratio increased gradually through the 1950s but at a more rapid pace in the early 1960s because of import substitution-induced investment. By the late 1960s foreign investment was 10 percent of the total, and still increasing gradually.

In a ratio of 5 to 1, foreign investment was direct and portfolio investment. The direct investment was in chemicals, cosmetics and paints, auto assembly, and food processing; and portfolio investment was in cement and chemicals. Foreign investment does not dominate industry. The small amounts, however, understate the importance of foreign investment since the most dynamic sectors receive the foreign investment, along with technology and managerial assistance, and associate themselves with foreign enterprise. The unenthusiastic welcome of foreign investment by the government has deterred but not stopped foreign investment.

Industry's capital stock quintupled between 1950 and 1969 (in 1957 prices) and between 1970 and 1974 more than doubled again (in 1968 prices). Capital stock in electric power increased the most rapidly; Bs217 million in 1950, Bs1,586 million in 1960, and Bs2,515 million in 1969, all in 1957 prices. From 1970 to 1974 accumulated net investment summed to Bs2,593, in 1968 prices. Capital stock in manufacturing was Bs1,536 million in 1950, Bs4,019 million in 1960, and Bs6,716 million in 1969, in 1957 prices. Between 1970 and 1974 capital accumulation in manufacturing was Bs6,921 million in 1968 prices. The price comparison problem notwithstanding, the last twenty-five years has witnessed the creation of a substantial industrial establishment in Venezuela.

The growth of capital stock and output in industry has been roughly parallel. The capital-output ratio for manufacturing has been almost constant at a little bit over one. In the 1950s the coefficient dipped, then increased in the early 1960s but there has been no trend. The capital-output ratio in electric power is higher and has more variation, but again, no overall trend. In the late 1950s while building Macagua I, the industry's coefficient increased, only to decline later with project completion and added output.

The amount of capital Venezuelan workers had at their disposal increased rapidly in the 1950s, but in the 1960s tended to decline before increasing again in the 1970s. The capital-intensive nature of the large hydroelectric facilities conferred on workers in that sector the greatest amount of capital, and construction had the least amount of capital per worker. The slower rates of investment in manufacturing, and its larger payrolls beginning in the 1960s, reduced the capital per worker ratio slightly.

Venezuela has employed its physical plant in industry more effectively over recent years. In 1967 the utilization ratio declined, but in 1968 and 1969 it was back up to nearly the 1966 level. It remained about the same in 1970, but in 1971 and subsequently it has increased, and at the end of 1973 was 74 percent. Basic metals, transport equipment, tobacco and tires made better use of capacity than such sectors as metal products, furniture and ready-made clothes.

Employment

At the beginning of 1970 the labor force in industry was 703,000 workers, an increase of 145 percent over 1950. The growth of labor demand by industry was sufficiently high that the proportion of labor in industry increased significantly. In 1950 all of industry employed 18 percent of the labor force; 12 percent in manufacturing, 6 percent in construction, and a minute fraction in electric power. Some shift occurred in the 1950s but the most pronounced increase in the industry share was in the 1960s. In 1969 manufacturing used 17 percent of the labor force, construction still 6 percent, and electric power 1 percent.

Output per worker in industry, however, has not improved significantly. In 1950 industrial output per worker was 95 percent of the output per worker for the entire economy. During the 1950s industrial output per worker outpaced the average, so that by 1959 the former exceeded the latter by 3 percent. In the 1960s, however, industrial output per worker stagnated with a tendency toward decline, especially in the late 1960s, even more so than average output per worker. In 1969 industrial output per worker fell short of the average by 21 percent.

The exception to this trend was in electric power production. Its output per worker was high and rising over the two decades and even registered more impressive gains in the 1960s than in the 1950s. Electric power output per worker rose 22 percent between 1950 and 1959, and 119 percent between 1960 and 1969. The electric power industry is highly capital intensive and employs relatively few and only specialized workers. In construction the trend in output per worker was upward through 1963, and then declined slightly.

The behavior of manufacturing output per worker was paradoxical. It increased more rapidly than the average for the economy in the 1950s, but during the 1960s the declines, although small, were systematic. By 1969 output per worker in manufacturing was only 90 percent of what it had been in 1950. In most industrialization processes the expectation is:

increasing output per worker as industry becomes more capital intensive. In the 1960s in Venezuela, manufacturing became more labor-intensive.

The explanations for declining output per worker in manufacturing are the lower rates of manufacturing investment in the 1960s, the import-substitute nature of the industrialization, the changing mix of industry, and government and union labor policies. Lower net investment rates resulted in a decline in the 1960s in the amount of capital per worker, and the rate of increase of capital stock slowed down. For example, in the 1950s capital stock increased 162 percent, and in the 1960s, 67 percent. Venezuelan industry became less capital-intensive and required more labor in production processes.

Manufacturing responded to the import substitution program by embarking upon production in which it was inefficient. With the lesser protection and boom atmosphere of most of the 1950s, manufacturing exhausted its growth potential without additional protection. When the protection appeared, industrial enterprises necessarily employed labor and other factors inefficiently. In addition, the mix of products changed so that the expanding industries of the 1960s used more labor than those of the 1950s. For example, in the 1950s, petroleum refining, which uses a very small labor force, expanded rapidly, but in the 1960s this industry's growth slowed down and other industries, such as chemicals, paper, automobiles and food processing, using more labor, expanded more rapidly in the 1960s.

In the 1960s unemployment was a serious economic and social problem, and the government, directly and through support of labor unions, encouraged the use of labor. At the same time, exports were not prospering and the government endeavored to limit imports, including capital goods imports. Government encouragement of the labor unions and enforcement of labor laws made it difficult and costly to discharge inefficient workers, shortened the work week, and discouraged employers from insisting on high and rising productivity. Finally, as the industrial labor force expanded in the 1970s, the new entries were probably not as productive as those employed earlier. Manufacturing dipped deeper and deeper into the labor force and came out increasingly with workers who did not have the basic aptitudes and skills needed in manufacturing.

Management

As industry has expanded in the last quarter century, the management of enterprise took on an increasingly professional look. The traditional family firm began to give way to hired professionals in positions of high responsibility; the one-man firm employed specialists to perform many functions. Enterprises introduced modern accounting techniques and new management methods. In the late 1950s and 1960s most universities established schools of economics and business administration, specialized courses in business proliferated, and the private sector formed a graduate school of business. While management does not measure up to the standards of the industrialized countries, or even some semi-industrialized countries such as Mexico, a clear trend toward modern management practices was manifest.

Managerial inefficiency, outdated practices, and reluctance to accept the harsh realities of modern industry still afflicts many enterprises, especially the small and medium size firms. Nepotism is frequent, partisan politics dictates personnel choices and decisions, who one knows and not what one can do determines placement, pay and advancement. The merit system has worked its way from complete absence to lipservice. Among many firms, low-volume, high-unit-profit operations take precedence over high-volume, low-unit-profit expansion, especially when the latter disturbs a comfortable routine. Management tends to avoid risk and commitments to do things it has not already done before. The drive, the ruthlessness, the energy and the determination of modern industrial enterprises are often lacking.

The supply of entrepreneurs in Venezuela has never been large. The social structure and cultural orientation, conditioned by centuries of Spanish tradition and agrarian activity, was not conducive to the development of the human resources necessary in industry. Contacts by Venezuelans with the advanced industrialized countries have been limited and not the kind designed to encourage the transfer of interest in industrial operations and technology.

The educational base at all levels has been slender. Before 1936 only a tiny fraction finished even sixth grade and only a handful achieved a university education. Education was oriented to law, medicine and the humanities, and educational methods emphasized rote learning. The practical arts, engineering, science and technology, and administration have come only in recent years. The man of letters, the landowner, the politician gained the prestige and the rewards.

The deficiencies in entrepreneurship and in the number and quality of management professionals began to dissipate in the 1960s. The government began to pour money into a serious effort to increase the size of the educational establishment and to modify its character to conform to the industrial and professional needs of industrial development. Higher incomes permitted greater contacts abroad, including contacts among technical and professional people associated with industry. The social and cultural image of industry improved and the economy began to reward merit and achievement to a greater extent.

In the 1950s a large influx of foreigners, mainly Europeans, played a role in initiating the industrial impulse. Between 1950 and 1959, 349,000 foreigners migrated to Venezuela, became professionals, businessmen, industrialists, skilled workers. By their actions, by example, and in combination with changing values, industry improved. Venezuelan society, as a result of the social, political and economic upheavals of the last quarter century, has now begun to generate its own supply of "new men," entrepreneurs and specialists needed to operate modern industry and stimulate its growth.

Commerce and Services

Commerce and services historically lag behind the development of extractive sectors and industry, following in the wake of these more dynamic sectors until a late stage of economic evolution. In industrially advanced, high income countries the tertiary sectors, especially services, experience rapid growth only after industry has matured and resource exploitation has settled down to a small but permanent niche in the economy. Consumers shift their attention from goods to services and make increasing demands on government, propelling the service sector forward rapidly.

Venezuela has not developed in this way. Commerce and services have for decades grown rapidly and since 1960 have been highly dynamic sectors. Their growth replaced the growth of petroleum in real terms in the 1960s and 1970s, and they have grown as fast as industry. In 1950 commerce and services were already 45 percent of gross territorial product. In the succeeding quarter century, their share has increased and is now well over one-half of the economy.

Even during the rapid petroleum and industrial growth of the 1950s, commerce and services almost kept us, growing only slightly less than the economy as a whole. Indeed, services alone did match the economy's growth. In the 1960s commerce and services registered growth half again as fast as that of the economy. In the 1970s they continued to grow more rapidly than the economy, but by not as wide a margin as in the previous decade when these sectors were the leading sectors.

The 1960s saw the blossoming of services especially. The prosperity of the upper and middle classes, industrial expansion, and changed government priorities set the stage. In 1961 alone, the service sector share of gross territorial product jumped from 26 percent to 29 percent. In that year the net growth of the economy was the same as the growth of the service sector. During the decade services grew almost twice as fast as the economy and accounted for nearly one-half of all economic growth in the decade.

Service growth continued in the 1970s. Between 1970 and 1974, for example, government services, as well as transportation and communications grew at double the rate of growth of the economy in real terms. Community, social and personal services grew even more rapidly. The share of the service sector in the economy increased 10 percent. Venezuela has developed a service economy, devoting in 1974, 43 percent of its output (in 1968 prices) to the service sector.

The reasons for the development of the Venezuelan service sector resides in the nature and manner of development of the country's resources. In already developed countries agriculture developed first and private landowners invested their accumulated wealth and profits in industry, which then became the leading sector, pulling services along behind it. To be sure, the government and private sector built the necessary infrastructure, but the people wanted goods and more goods, and industry provided them. Only at high income levels, and when almost satiated with goods did consumers switch to services, promoting their growth even beyond that of industry.

Venezuela moved directly from primary to tertiary emphasis in production and employment, all but omitting the stage of industry as the leading sector. While industry has grown and become more important, much of the real decline in extractive production has been replaced by service production. Venezuela's extractive sector is petroleum, not agriculture, and petroleum's development conferred benefits on foreigners and the government, rather than owners of agricultural land.

The petroleum companies used their share of the extractive sector surplus to develop foreign economies, notably the United States, not Venezuelan industry. Even the foreign and backward linkages were weak, promoting significant growth only in the petroleum refining industry, which in real terms has grown slowly in recent years. The government share of the surplus of the extractive sector also did not build industry. The government confined its industrial activities to a relatively few projects, leaving most of industrial production as the preserve of the private sector.

Rather, government expenditures based on the petroleum surplus built economic and social infrastructure and provided enlarged services, compelling the rapid growth of the service sector. But industry, constrained by lack of experience and inefficiency, grew as incompletely protected markets, financial and technical assistance, and economic infrastructure grew. Service growth stimulated industrial growth, rather than the other way around.

The nature of the political system determined the relative growth of services and industry. Technically, the new government had the option of developing industry rapidly under public auspices. To have done so, however, would have alienated the private economic sector on whom the fledgling democratic government had to rely for survival. On the basis of this political reality, the government chose to spend its petroleum money on services, expecting thereby not only to improve social and economic conditions, but also to stimulate private industrial growth. It did, but not by enough for industry to move into the position of the dominant leading sector of the economy.

Summary

While Venezuela has not become an industrial country in the last quarter century, it did build the foundation for the industrialization process. Industry grew more rapidly than the rest of the economy in production and employment. It branched out into new fields and expanded production in all lines. Government policy, through protection and trade policy, as well as financial and other measures, promoted industrial growth. Except in large scale, resource-related industries, however, the government did not participate in production. Nationalization includes petroleum refining, and plans for the use of reserve funds indicate greater government production, but still confined to resource industries.

Inefficiency and small markets have hampered industrial growth. Debilitating protection, insufficient investment, low productivity labor, and inadequate management bred the inefficiency and unequal income distribution and kept markets small. Industry did make progress in reducing the effects of some of the deficiencies, but not sufficient to compete and to expand without government support. So far, Venezuela has made only the beginning of the industrial transformation.

Services, not industry, replaced petroleum as the dynamic sector in the 1960s and Venezuela has become increasingly a service economy. The survival of democracy required that industry develop predominantly as private enterprise, leaving the government the task of stimulating industrial growth indirectly through the expansion of services and infrastructure.

NOTES

1. The data on industry come from publications of the Banco Central de Venezuela, and the *Anuario Estadístico*.
2. M. A. Falcón Urbano. *Desarrollo e Industrialización de Venezuela*, 1969, pp. 163–190.
3. *Un Programa para Venezuela*, Caracas: Associación Pro-Venezuela, 1971, pp. 29–92.
4. Orlando Araujo. *Situación Industrial de Venezuela*, 1969, pp. 28–61.
5. Tomás Enrique Carrillo Batalla. *El Desarrollo del Sector Manufacturero Industrial en la Economía Venezolana*, 1962.
6. *Cuarto Menesaje del Presidente Rafael Caldera*, Caracas: Oficina Central de Información, 1973, pp. 217–250; *Primer Mensaje del Presidente Carlos Andrés Pérez*, Caracas: Oficina Central de Información, 1975, pp. 199–251; and other presidential messages.
7. "Venezuela," special section of *The Financial Times*, London: 17 November 1975, pp. 13–32.
8. Domingo Alberto Rangel. *La Oligarquía del Dinero*, 1972, pp. 126–127.
9. *Primer Mensaje del Presidente Carlos Andrés Pérez*, 1975, pp. 206–211.

Chapter XIV

The Venezuelan Transformation

The nationalization of the $6 billion petroleum industry at the end of 1975 exemplifies the transformation of the Venezuelan national economy and politics in the last forty years. The economy has grown at a vertiginous rate, output nearly doubling every decade since 1935, and its structure has changed. In addition, Venezuela has established and maintained a representative democracy for more than a decade and a half after a century of dictators.

The economic transformation stems from petroleum plus innovative, nationalistic petroleum policies, coupled with economic policies that have helped to develop and diversify the economy and to distribute its benefits more equitably. The political transformation reflects the taming of the military and the installation of a party-based political system that can resolve conflicts pacifically. The circumstances, events and policies of Venezuela are unique and represent a new and different politico-economic development pattern.

The Political System

Before 1936 Venezuelans had no experience with representative self-government. In 125 years of independence dictators had ruled the country under a score of constitutions. General Juan Vicente Gómez had run Venezuela from 1908 to 1935 as his personal ranch and had ruthlessly filled his jails with political enemies. His regime was so brutal that even today no street or monument bears his name. When he died in 1935 the people jubilantly took to the street and forced the government to confiscate his properties and oust his clan.

The successor to Gómez was his war minister, General Eleazar López Contreras, a man of a different stripe. While his regime was authoritarian, it was also more mellow, permitted more press freedom, political gatherings and public opposition. Under López, Venezuela began to taste democracy, but within strict limits and without electoral freedom. At the end of his term in 1941, López chose his successor, General Isaías Medina Angarita. Medina was even more liberal, permitted the formation of politicial parties, and introduced economic and political reforms. The trend under López and Medina was toward more democracy, but still under military tutelage.

The period from 1945 to 1948 was a short burst of political energy and an experiment in more representative goverment. For long years the civilian opposition had waited in the wings. In 1945 it had its chance, but only with the help of disaffected military officers. The coalition deposed Medina, established a civilian government, and embarked on political and economic reform. By today's standards the government was not radical, but at that time, and in Venezuela, it was a major departure. Rómulo Betancourt was the interim president and Rómulo Gallegos, novelist and educator, succeeded him in 1947 in Venezuela's first popular election.

The hasty reforms and boisterous partisan politics were too much for the military and in 1948 it reclaimed the government. For nine more years the military governed Venezuela, at first through a junta, and from 1952 to 1958 through the dictatorship of General Marcos Pérez Jiménez. Political freedom vanished and economic and political irresponsibility reigned. The dictatorship did, however, provide Venezuelans with an opportunity to compare dictatorship with respresentative government, and in the scales democracy won.

By 1958 the abuses were so obvious that all segments of Venezuelan society rose up to remove the dictator in favor of a junta that promised free elections. Political parties reappeared and conducted a spirited campaign after agreeing to abide by its results with the winner forming a coalition government. The military kept its pledge and Betancourt again became president as head of Acción Democrática party. Although beleaguered by terrorists and economic recession, his multi-party government effected many reforms and, contrary to the expectations of most, finished its term. In 1964, for the first time, a popularly elected president surrendered the tri-colored sash to his popularly elected successor.

President Raúl Leoni, also of Acción Democrática, continued the policies of Betancourt and introduced new reforms. Congress increased its participation in decision-making. Violence subsided and the economy recovered. The party system held together and the parties matured. In [illegible] Venezuela registered another first: Rafael Caldera of the opposition

Social Christian party (COPEI) won the election and became president. In 1974 Acción Democrática came back to power under Carlos Andrés Pérez in an election in which the two main parties polled 85 percent of the vote despite a dozen other candidates.

The secret of Venezuelan representative government is the evolution of political rules of the game. The first rule says that the maintenance of democracy is more important than any party, personality, or policy; better defeat and a chance for comeback than a victory which undermines the political system. All parties also agree that political activity must be channelled through the parties: no more street mobs, student violence, or appeals to the military. Neither winners nor losers are right. Winners have the privilege of pursuing their policies and programs; losers have the privilege of opposing them. The government, even though of a single party, governs for all the people.

Four free elections, three constitutional transitions, two party transfers, and the emergence of a two-party system are evidence of the stability of the Venezuelan political system. Congress plays a vital role in policy formulation and controls the budget. The press is free. All parties vie with one another in nationalistic fervor and the two main parties are middle-of-the-road coalitions, under constant attack by the noisy left and sullen right. In forty years Venezuelan politics have changed from absolute dictatorship, through an uncertain experiment in representative government, to a robust party-based democracy.

Economic Growth and Structure

In 1935 Venezuela was a backward agricultural country with a petroleum enclave primarily of benefit to foreigners. Most Venezuelans eked out a mean living from the soil. Per capita national income was less than $150. The productivity of agriculture was pitifully low and almost no industry existed. The country had changed little from colonial days. Only the favorites of the dictator prospered. None of the elements to promote economic growth, except oil, seemed to be present. But petroleum, combined with nationalistic policies, was enough.

Based on the manipulation of petroleum, the Venezuelan economy has experienced real growth of seven percent per year for the last forty years. Per capita income in 1974 was over $2,000, fourteen times what it was in 1935. The economy is booming, Treasury coffers are full, and for the first time in its history, Venezuela is experiencing double digit inflation. The 1973–1974 quadrupling of petroleum prices has brought in more money than the economy can absorb. Even should the price of petroleum sag, economic prospects are bright.

Growth has not always been smooth, since it depended on petroleum which has been one-fourth or more of total product and nearly all of exports. When petroleum demand rode high, the economy zipped along, as in the late 1940s and 1950s. In the early 1960s a recession staggered the economy and throughout the decade exports stagnated and petroleum prices declined. Even so, the growth rate for the decade for the economy was six percent per year. High prosperity began to return in 1972. Then those incredible years of 1973 when gross territorial product increased 21 percent, and 1974 when it leaped ahead 66 percent.

High rates of public and private investment, both made possible by petroleum, have propelled the economy forward. The $20 million trickle of revenues of 1935 became a river in the 1940s, a flood in the 1950s and 1960s, and a deluge in the 1970s. In 1974 public revenue was $10 billion. These funds have been used to build social capital in housing, transportation and education, as well as productive capital in agriculture and industry. Much of the vigorous private sector growth originates from public expenditures as they filter through the economy.

Growth has altered the economic landscape. Petroleum production reached its zenith in 1970; the government has cut back production since then. Petroleum income, based on high prices, supports the government, which now dominates economic life. It owns and operates a steel mill, petrochemical plants, hydroelectric facilities (including one of the world's largest at Guri), an aluminum plant, and other enterprises. Natural gas, iron ore and other mining operations have been nationalized, and with the nationalization of petroleum, the government will produce well over one-half of the economy's product.

The private sector dominates agriculture, much of industry, commerce, and services. Agricultural productivity has increased and more land is in use with fewer workers. The agrarian reform program has stimulated output by placing more land in the hands of the rural poor. Agricultural production has grown only slightly less than output as a whole. An import-substitution program, relying on high tariffs, as well as government credit, has promoted industrial growth. Industry is an increasing share of output—25 percent in 1974—and Venezuela is now a semi-industrialized country. Services, nourished by government spending, is one of the most rapidly growing sectors. But still, everything depends on the government and the government depends on petroleum.

The distribution of income among economic classes has improved. A relatively few still receive most of the income, but the lot of the growing middle class and even the poor is advancing. Foreigners get less and regressive taxes are declining; minimum wages, maximum prices on goods of primary necessity, and more employment opportunities help low income families. In 1974, for example, the government decreed that all elevators and washrooms had to have attendants, opening up thousands of new jobs for the poor. Public funds poured into education, medical facilities, social security, low-cost housing, sanitation and social services which benefit the poor more than the rich.

Petroleum Policy

In 1935 petroleum policy served the foreign petroleum companies. These companies did what they wanted, wrote the petroleum laws, and paid little in taxes. Money interested the government—the dictator—but the small amount paid by the companies was so great compared to previous revenues that the government paid off all debts and ran surpluses. After the death of the dictator, the intense public petroleum debate led to higher taxes and more regulation of the industry. The outlines of the new policy emerged: maximize the public yield of petroleum through high prices and taxes, conservation and regulation, and eventual takeover of the industry, which in the meantime would function as a public utility, profitable to the government.

Two major moves came in the early 1940s. The government passed a progressive income tax law, the first of its kind among less developed petroleum exporting countries. In addition, in 1943 a new petroleum law cancelled all previous concessions, sanctioned under a dozen previous laws—some very generous to the companies—and replaced them with 40-year nonrenewable concessions having uniform and stricter conditions and higher royalities. These two measures were key elements in the plan to convert petroleum into the engine of Venezuelan economic growth.

But the plan had a flaw in it. In a vertically integrated industry such as petroleum the prices charged for primary and intermediate products in different places, all under the control of a single enterprise, can be manipulated to benefit the company. In addition, the world petroleum price system is highly imperfect, subject to management by a few companies. The companies, in order to minimize payments to the government, chose low prices in calculating their tax and royalty obligations. Early efforts of the government to raise prices had little effect.

In 1948 the government attacked this price problem frontally. The petroleum law reserved to the government the option of receiving its royalty in crude oil. In a surprise move, the government decided to take crude instead of cash and sold it on the market. The companies, reluctant to have the government in their business, agreed to pay royalties based on a higher crude oil price. It was a start, but the companies still decided the price.

Petroleum reforms were dramatic during the 1945–1948 democratic regime. The emerging policy, in addition to price and tax elements, included conservation and environmental protection measures, field inspections, and the training of Venezuelans to staff positions in the industry and government. The government decided to grant no more concessions and planned a state-owned petroleum company. In its final act, introduced weeks before the military took over again in 1948, the government split evenly petroleum net income. In a few years this rule became standard throughout the world.

In 1958 the government renewed active intervention in the industry after a decade in which the military regime introduced no new measures but did grant more concessions. The state-owned enterprise became a reality in 1960, and by 1975, though still small, it was the fifth largest petroleum enterprise in the country. The government initiated service contracts to replace the concession system. It also tightened regulation and supervision, increased taxes, and in 1958 the obligatory government share became two-thirds. By 1975 the government was taking 85 percent of petroleum profits. Venezuelans have taken over all but the top management jobs and highly specialized professional and executive tasks. On the eve of nationalization only 500 foreigners in a labor force of more than 20,000 work in the industry.

The government also renewed the attack on the price problem. It sued the companies for back taxes, alleging underpricing in tax calculations. Long and difficult negotiations ensued. The long-standing policy of talking continuously with the companies paid off. They paid some back taxes, but more important, the companies agreed in 1967 to pay taxes based on a negotiated reference price, not market price. Reference prices reflected what the government thought prices should be and were higher than market prices. In 1971 the government began to set reference prices unilaterally.

Reference prices were the key to the control of the price of petroleum. If the government can determine the price with which the companies calculate income for tax purposes (and taxes are a significant element of costs), then the companies must set a price which covers not only non-tax costs and profits, but also taxes computed with reference prices. By changing the reference price the government can force the companies to change the market price. The higher petroleum price, however, does not squeeze many buyers out of the market, so the value of petroleum sales, as well as government revenues, go up. The government, by controlling reference prices indirectly, controls market prices and its own revenues, and the companies become commissioned intermediaries.

Still there is a missing link. No single country can determine the market price of petroleum. If Venezuela tried to set a price much higher than that of other exporters, the oil companies would cut production there and expand where prices are lower. Venezuelan eagerness for more revenues indeed had this result. The control of prices can be taken away from the companies only if most of the major exporters set about the same price and then sell only the amount which clears the market.

Venezuela was an early proponent of international cooperation. It promoted meetings of oil exporters and in 1960 was a founding member of the Organization of Petroleum Exporting Countries. Indeed, the Venezuelan oil minister, Dr. Juan Pablo Pérez Alfonzo, and his Saudi Arabian counterpart, Sheikh Abdullah Tariki, were the chief architects of his bold plan to control world oil. Once OPEC was a reality, Venezula successfully persuaded many other exporters to adopt policies similar to its own. In the first decade of OPEC, membership expanded to include most large exporters, and members learned the oil business and installed uniform policies.

In the last fifteen years Venezuela has led the way for OPEC members in demanding and getting more oil revenues and influence in the world petroleum industry. In 1973, when the oil crisis came, OPEC was ready. It wrested control of the price of oil from the private cartel and began to operate as an intergovernmental cartel. Venezuela remains a faithful member of OPEC, and although it did not join in the 1973 oil embargo, it is one of the countries which continues to press for higher prices.

The nationalization of petroleum at the end of 1975 was politically inspired, a result of intense nationalistic pressure. In 1974 the government had again increased tax rates so that even without nationalization more than 95 percent of petroleum income returns to the Venezuelan economy. Furthermore, the industry would have reverted to the government beginning in 1983 in any case, without the payment of nearly $1 billion to the oil companies. But the political pressure cooker demanded attention and only nationalization could release its energies.

Even so, nationalization is probably not premature. A non-political holding company, Petróleos de Venezuela (Petroven), ninth largest petroleum company in the world and among the 50 largest enterprises in the world, will operate the existing companies as public corporations. MIT-trained General Rafael Alfonso Ravard, one of Venezuela's most able administrators, will run Petroven free from political interference. The nationalization was orderly and peaceful; new logos have replaced the old company symbols but most people stayed on the job.

The nationalization provides for continued cooperation with private companies in technology and distribution. Petroven, for the time being, will market its oil through the international companies through contractual arrangements. These companies will also provide technological assistance, especially as new technology develops. Venezuela has the financial and material resources, the technical and managerial capability and experience, a secure U.S. market, a framework of international cooperation, and a sound economy. The patient and probing petroleum policy has paid off.

Economic Policy

Venezuela has found out that it is not enough just to have money. The money has to be spent efficiently and productively. But, in order to spend the money usefully, difficult choices must be made among competing uses. Before 1936 no government economic policy existed. As the personal property of the dictator, the government spent money to suit his goals, including his own enrichment, as well as the security and stability of his regime.

The famous slogan of Arturo Uslar Pietri, now Venezuela's ambassador to UNESCO, summarizes government economic policy. Venezuela must "sow the petroleum," he said in 1936. The petroleum money must be spent to eliminate the need for petroleum money. Roads, sewers, hospitals, industrial parks, schools, houses, airlines, factories, irrigation, banks social services, ports, telephones—these and many other capital improvements were lacking. Bit by bit over the last forty years many have appeared. And more will appear as the petroleum money rolls in.

In those activities where the private economic sector saw the prospect of profit from government orders, credit, expenditure-induced expansion, high tariffs, the government encouraged private business to build and expand. In unprofitable sectors or where the job was too big or risky, or vital to the national interest, the government assumed responsibility. Among the latter are steel, hydroelectric power, aluminum, petrochemicals; all resource-related basic industry. Unbridled capitalism has little scope; the government regulates and taxes the private sector and supports strong labor unions. Socialist thinking plays no role and indeed socialism has repeatedly been rejected. Economic policy has been pragmatic and experimental.

In 1960 the government introduced systematic economic planning. The emphasis, however, is on coordination within the government and making arrangements in which private sector and government activities mesh. Cost-benefit studies guide public expenditures and budgeting-by-program spurs government productivity. The plan is not important; the process of planning and coordination is the center piece, not only in the small planning agency, but also in every ministry and institute. The planners keep the pressure up but their powers lie in education, persuasion and coordination, not in edicts.

Financial policy is conservative, although the financial community has expanded frantically in the last two decades. A strong government-owned central bank controls not only banks and other financial intermediaries, which reach everywhere in the country, but also many specialized government lending institutions. The government monitors foreign investment closely and reserves banking and finance to Venezuelans. In the past, finance served only the oligarchy; today, every financial service is available to a widening circle of clients. Venezuelan bankers are beginning to branch out into international finance and Caracas is becoming a major financial center.

Conservatism also marks international economic policy. Acutely aware of its dependence on petroleum exports, Venezuela is trying to promote other exports and to keep imports down. High tariffs, recently revised to bring them into line with the practices of industrialized countries, save foreign exchange and promote industrial and agricultural expansion. International debt is small, even including the nationalization debt, which will be paid in ten years. The President could write a check any day for the entire foreign debt and still have plenty left over. The bolívar, worth a little less than a quarter, has always been a hard currency and is now backed by $8.8 billion in international reserves.

Leadership

No leadership was expected of, or came from Venezuela before 1936. Its claim to fame outside its borders was as the stereotype of the strutting and repressive military dictator. Since then Venezuela has risen to a position of continental and world leadership, not only by example, but also by positive assertion of its influence. The international coming of age of Venezuela reflects its resource policies, economic progress and policies, and democratic political system. Today the world respects Venezuela as one of the most responsible and vigorous spokesmen for the less developed countries.

Venezuela has shown that nationalistic policies directed at the foreign-owned leading sector which dominated the economy can be implemented through negotiations, in a democratic setting, without disrupting the economy, and with growing benefits to the people. It has demonstrated also that the government, without the shibboleth of socialism, can develop the economy with pragmatic policies. This has given Venezuela the voice of moral authority even among countries not prepared to follow its example in all respects. Many primary producing countries look to Venezuelan policies, and pattern their own after them. Venezuelan responsibility and fairness toward the foreign private sector has earned the respect of industrialized countries and their private sectors.

Venezuela has established that a democratic party-based system can work in Latin America. Many have despaired, even to the point of arguing that democracy cannot function in less developed countries. In Venezuela it is working. A system of conflict resolution and policies tailored to the country's needs and capacity have created the most successful Latin American democracy. A comparison of Venezuela with authoritarian Cuba, chaotic Argentina, or repressed Chile is not lost on Venezuelans or others.

The present petroleum bonanza has given Venezuela another springboard for influence in the world. The new funds not in use developing the country are in international capital markets, making Venezuela a leader in recycling financial resources to petroleum consuming countries. Venezuela also lends money to Latin American countries hard hit by the high petroleum price. Its contributions to the International Monetary Fund, World Bank, and Interamerican Development Bank, as well as other international institutions, have increased. Venezuela has pledged money to the OPEC fund to help less developed countries. In 1975 Venezuelan officials chaired the annual meetings of the Fund and Bank. Wealth, progress and democracy are making Caracas a stopping place for business and political leaders from all over the world.

Several initiatives mark Venezuela's present international economic posture. The country is in the forefront of those who seek price increases for all raw materials, not just petroleum alone. It is encouraging the formation of OPEC-like organizations for other products. The establishment of link between raw material and industrial prices is another primary goal. With the leverage of petroleum, Venezuela and other countries seek to discuss new economic arrangements with industrial countries. In this new international economic order raw material producers would share more equitably in the benefits which come from their resources.

In Latin America Venezuela is an emerging continental leader. Its loans and financial assistance, its espousal of supportive policies, and its success have earned Venezuela the respect of the other countries. Along with Mexico, Venezuela is promoting the Latin American Economic System.

In this new arrangement, still in process of formation, Latin American countries, including Cuba, will coordinate their economic policies, undertake joint ventures, help one another, and from a position of collective strength will demand that the industrial countries share their economic and technological progress.

Shortcomings

Let it not be thought that no blemishes appear in the developments in Venezuela in recent years. While the role of the private oil companies has changed, dependence on petroleum continues; government income and the progress of the economy hang on decisions in consuming and other exporting countries. Venezuela has had little experience in marketing petroleum, and despite progress in training professionals and technicians, technology still emanates from the large oil companies. Venezuela will buy technology and will sell oil to the same companies which formerly produced the oil.

In spite of the diversity of the economy, it is still lopsided with most exports and a large chunk of total product originating in the petroleum sector. The industrial sector still produces mainly consumer goods; capital and intermediate goods must be imported. Progress in agriculture has not increased productivity enough to eliminate critically important food imports. Another disconcerting trend is the enlargement of the commercial and service sectors where productivity gains are difficult to achieve. The majority of Venezuelans are still poor and the list of unfinished tasks is very long.

The efficiency of the private sector needs improvement. The traditional outlook—low volume, high unit profits, limited competition—has bred few entrepreneurs, business timidity, and a willingness to depend on the government. Large interlocking enterprises dominate much of the private economy. Shielded by protective tariffs and operating in imperfectly competitive markets, Venezuelan business often feels small need to improve productivity or to take changes. Imports undersell Venezuelan products and exports other than petroleum have not penetrated foreign markets.

The Venezuelan government is a huge bureaucracy that would set no records for efficiency. It is sometimes used as a welfare program for the less fortunate party faithful. The government wastes vast sums on projects which cost too much or should not have been undertaken. Neither dictatorship nor democracy have guaranteed honesty or efficiency in government and the checks and balances sometimes do not work. A constant preoccupation of the leadership is waste, lack of discipline and motivation, and the tendency to look for the easy way.

Politics is inextricably bound with economics. The government undertakes a project because a politician favors it; a ministry fires a highly qualified technician in order to make way for a politically more reliable worker. Nothing happens unless it has the support of a key political figure. Much of the time and attention of the government goes to political in-fighting, with everything geared to the next election. The year before the election the government falters; during election year, campaigning paralyzes the government. Criticism fills the air; backbiting, refusal to compromise, and disdain for the point of view of others often characterizes the attitude of the government and political parties.

A serious impediment to development is the Venezuelan self-image. Venezuelans often tend to be incurably pessimistic. In the midst of prosperity, they think the economy is falling apart. The North Americans, the oligarchs, the politicians will wreck all hopes, steal all the money, turn every success into failure. With one of the best planning systems in Latin America, Venezuelans believe they cannot plan. Politicians periodically proclaim the demise of democracy and the collapse of the economy. Despite the evidence, few Venezuelans herald a great destiny for their country.

These shortcomings are not unique to Venezuela and indeed are less crippling than in most countries. They reflect immaturity, lack of experience, and hurry. Most of the deficiencies are self-correcting and many improvements can already be observed in economic and political life. The educational program, from primary grades to graduate schools of business, engineering, agronomy and petroleum economics is boosting productivity, upgrading skills, and making continued development almost certain. The new man, who wants progress for himself and his country, is increasingly in charge.

The transformation of the Venezuelan political and economic system in the last forty years compares favorably to the "miracles" of Germany and Japan. Venezuela will be one of the first less developed countries to attain developed-country status, only a few years from now. In learning to control its own resources, in developing its own brand of representative government, Venezuela has pioneered a new path.

BIBLIOGRAPHY

Abercrombie, Thomas. "Venezuela Builds on Oil," *National Geographic,* CXXIII, March 1963, 344–387.

Acedo Mendoza, Carlos. *Venezuela Ruta y Destino,* Caracas: Editorial Fondo Común, 1971.

Acosta H., Eduardo. *La Comisión Económica de la OPEP*, Caracas: Editorial Arte, 1971.

———. *Este Petróleo Es Venezolano*, Caracas: Editorial Arte, 1970.

———. *Fundamentos de una Política Nacional para Venezuela*, Mérida: Universidad de Los Andes, 1967.

Akins, James. "The Oil Crisis: This Time the Wolf Is Here," *Foreign Affairs,* Vol. 51, No. 3, Spring 1973, 462–490.

Alamo, Roberto. *Crecimiento, Urbanización, y Estructuras Demográficas*, Caracas: IDES, 1964.

Alexander, Robert. *The Communist Party of Venezuela*, Stanford: Hoover Institution Press, 1969.

———. *Prophets of the Revolution,* New York: Macmillan, 1962.

———. *The Venezuelan Democratic Revolution,* New Brunswick: Rutgers University Press, 1964.

Allen, Robert Loring. *Lecciones de Economía Regional,* Mérida: Universidad de Los Andes, 1972.

———. "Regional Development and Public Finance in Venezuela," in *Public Finance and Welfare,* Paul Kleinsorge, ed., Eugene: University of Oregon Books, 1966.

Anzala, Eloy. *Función del Capital Extranjero en el Desarrollo de Venezuela*, Caracas: 1965.

Araujo, Orlando. *Situación Industrial de Venezuela,* Caracas: Universidad Central de Venezuela, 1969.

Arcila Farías, Eduardo. *Economía Colonial de Venezuela,* 2a Edición, Caracas: Italgráfica, 1973.

———. *El Régimen de la Encomienda en Venezuela,* 2a Edición, Caracas: Universidad Central de Venezuela, 1966.

Arraíz, Antonio. *Geografía Económica de Venezuela,* Caracas: Cultural Venezolana, 1956.

Arrellano Moreno, Antonio. *Fuentes para la Historia Económica de Venezuela,* Caracas: Tercera Conferencia Internacional de la Agricultura, 1951.

———. *Guía de Historia de Venezuela, 1498–1968,* Caracas: Síntesis Dosmil, 1971.

———. *Orígenes de la Economía Venezolana,* 2a edición, Caracas: Ediciones Edime, 1960.

———. *Temas de Economía,* Caracas: Imprenta Nacional, 1970.

Asociación Pro-Venezuela. *Un Programa para Venezuela,* Caracas: Pro-Venezuela, 1971.

Balestrini, César. *Economía Minera y Petrolera,* Caracas: Universidad Central de Venezuela, 1959.

———. *La Industria del Mineral de Hierro en Venezuela,* Caracas: Ediciones del Cuatricentenario de Caracas, 1967.

———. *La Industria Petrolera en América Latina,* Caracas: Universidad Central de Venezuela, 1971.

———. *La Industria Petrolera en Venezuela y el Cuatricentenario de Caracas,* Caracas: Ediciones del Cuatricentenario de Caracas, 1966.

———. *Los Precios del Petróleo y la Participación Fiscal de Venezuela,* Caracas: Universidad Central de Venezuela, 1974.

Baloyra, Enrique A. "Oil Policies and Budgets in Venezuela," *Latin American Research Review,* Vol. IX, No. 2., Summer 1974, pp. 28–72.

Banco Central de Venezuela. *La Economía Venezolana en los Ultimos Vienticinco Años,* Caracas: Banco Central de Venezuela, 1966.

———. *La Economía Venezolana en los Ultimos Treinta Años,* Caracas: Banco Central de Venezuela, 1971.

———. *Ingreso Nacional de Venezuela,* Caracas: Editorial Relámpago, 1949.

———. *Memorias e Informe,* Caracas: 1940– .

———. *Revista,* Caracas: 1941– .

———. *Síntesis de la Economía Venezolana, 1961–1967,* Caracas: Banco Central de Venezuela, 1968.

Bendahan, Daniel. *La Legislación Venezolana sobre Hidrocarburos,* Caracas: Empresa El Cojo, 1969.

Bernstein, Harry. *Venezuela and Colombia,* Englewood Cliffs: Prentice-Hall, 1964.

Betancourt, Rómulo. *Pensamiento y Acción,* México: Fondo de Cultura Económica, 1950.

———. *Problemas Venezolanos,* Santiago de Chile: Editorial Futuro, 1940.

———. *La Revolución Democrática en Venezuela,* Caracas: Imprenta Nacional, 1968.

———. *Venezuela: Política y Petróleo,* México: Fondo de Cultura Económica, 1956.

Blank, David Eugene. *Politics in Venezuela,* Boston: Little Brown and Company, 1973.

Bonilla, Frank and Silva Michelena, José. *Studying the Venezuelan Polity,* Cambridge: MIT Press, 1968.

———. *The Politics of Change in Venezuela,* Cambridge: MIT Press, 1967.

Brito Figueroa, Federico. *Ensayos de Historia Social Venezolana,* Caracas: Universidad Central de Venezuela, 1960.

———. *La Estructura Social y Deomgráfica de Venezuela Colonial,* Ediciones Historia, Caracas: Universidad Central de Venezuela, 1961.

———. *Historia Económica y Social de Venezuela,* Caracas: Universidad Central de Venezuela, 1966.

———. *Venezuela Contemporánea: País Colonial?,* Caracas: Venediciones, 1972.

———. *Venezuela Siglo XX,* Habana: Casa de las Américas, 1967.

Burggraaff, Winfield. *The Venezuelan Armed Forces in Politics, 1935–1959,* Columbia: University of Missouri Press, 1972.

Carrillo Batalla, Tomás Enrique. *El Desarrollo del Sector Manufacturero Industrial de la Economía Venzolana,* Caracas: Universidad Central de Venezuela, 1972.

———. *La Economía del Comercio Internacional de Venezuela,* Caracas: Editorial Mundo Económico, 1962.

———. *Moneda, Crédito y Banca en Venezuela,* Caracas: Banco Central de Venezuela, 1964.

Castillo, Domingo B. *La Cuestión Monetaria en Venezuela,* Caracas: Ediciones de las Presidencia de la República, 1962.

Celis, Julio Paez. *Movimiento Migratorio del Area Metropolitana de Caracas,* Caracas: Universidad Central de Venezuela, 1960.

Chaves, Luís Fernando. *Geografía Agraria de Venezuela,* Caracas: Universidad Central de Venezuela, 1963.

Chen, Chi-Yi. *Economía Social de Trabajo: Caso de Venezuela,* Caracas: Universidad Católica de Andrés Bello, 1969.

———. *Estrategia del Desarrollo Regional: Caso de Venezuela,* Caracas: Editorial Arte, 1967.

———. *Los Movimientos Migratorios en Venezuela,* Caracas: Universidad Católica de Andrés Bello, 1967.

———. *Los Pobladores de Caracas y su Procedencia,* Caracas: Universidad Católica de Andrés Bello, 1970.

Colegio de Economistas de Venezuela. *Diagnóstico de la Economía de Venezuela,* Caracas: Universidad Central de Venezuela, 1964.

Córdoba, Armando. *Inversiones Extranjeras y Subdesarrollo,* Caracas: Universidad Central de Venezuela, 1973.

Córdoba, Armando y Silva Michelena, Hector. *Aspectos Técnicos del Subdesarrollo,* Caracas: Universidad Central de Venezuela, 1967.

Crazut, Rafael. *Consideraciones acerca de las Inversiones Privadas Extranajeras en Venezuela,* Caracas: Ediciones del Cuatricentenario de Caracas, 1967.

———. *El Banco Central de Venezuela,* Caracas: Banco Central de Venezuela, 1970.

Crist, Raymond E. *Venezuela,* New York: Doubleday, 1959.

Crist, Raymond E. and Leahy, Edward P., *Venezuela: Search for a Middle Ground,* New York: Van Nostrand Reinhold, 1969.

D'Ascoli, Carlos A. *Esquema Histórico-económico de Venezuela,* Caracas: Banco Central de Venezuela, 1970.

———. *Moneda, Comercio Exterior, y Otros Ensayos,* Caracas: Banco Central de Venezuela, 1967.

Deltec Panamérica. *El Mercado de Capitales en Venezuela,* México: Central de Estudios Monetarios Latinoamericanos, 1968.

Economic Commission for Latin America. "Economic Development in Venezuela in the 1950s," *Economic Bulletin for Latin America,* Vol. V, No. 1, March 1960, pp. 21–61.

Espinosa, Alfonso. *El Proceso Monetario: Venezuela 1930–1960,* Caracas: Editorial Arte, 1963.

Falcón Urbano, Miguel A. *Desarrollo e Industrialización de Venezuela,* Caracas: Universidad Central de Venezuela, 1969.

———. *Nuestro Petróleo, Riqueza Instrumental del Desarrollo en Venezuela,* Caracas: Editoria Didascalia, 1962.

Farley, Rawley. "The Economics of Realism: A Case Study of Economic Change in Venezuela," *The Economics of Latin America,* New York: Harper and Row, 1972, pp. 179–303.

First National City Bank. *Venezuela: An Economic Study,* New York: First National City Bank, 1966.

Friedman, John. *Economic Growth and Structure in Venezuela,* Caracas: Cuadernos de la Sociedad Venezolana de Planificación, 1963; *Revista de Economía Latinoamericana,* No. 6, 1962, pp. 115–204.

———. *Regional Development Policy: A Case Study of Venezuela,* Cambridge: MIT Press, 1966.

———. *Venezuela: From Doctrine to Dialogue,* Syracuse: Syracuse University Press, 1965.

Gallegos Ortíz, Rafael. *La Historia Política de Venezuela,* Caracas: Fundación de Cultura Universitaria, 1960.

Gilmore, Robert L. *Caudillism and Militarism in Venezuela,* Athens: Ohio University Press, 1964.

Gómez, Henry. *La Industria del Mineral de Hierro en Venezuela,* Caracas: Universidad Católica de Andrés Bello, 1970.

González Gorrondona, José Joaquin, hijo. *La Planificación Económica y su Aplicabilidad al Caso Venezolano,* Caracas: Ediciones del Cautricentenario de Caracas, 1967.

Hassan, Mostafa F. "Capital Flight in Venezuela, 1958–1960;" *Inter-American Economic Affairs,* Vol. XVII, Autumn 1963, pp. 53–73.

———. *Crecimiento Económico y Problemas de Empleo en Venezuela,* Caracas: Banco Central de Venezuela, 1973.

———. "High Growth, Unemployment and Planning in Venezuela," *Economic Development and Cultural Change,* January 1970.

———. "The Second Four-Year Plan of Venezuela," *Journal of Inter-American Studies,* Vol. 9, April 1967, pp. 296–320.

———. *Economic Growth and Employment Problems in Venezuela,* New York: Praeger, 1975.

Heaton, Louis E. *The Agricultural Development of Venezuela,* New York: Praeger, 1969.

Herrera Mendoza, J. M. *Historia de las Actuaciones del Banco Central de Venezuela,* Caracas: Empresa El Cojo, 1960.

Hill, George W. and Silva, José M. and Hill, Ruth. *La Vida Rural en Venezuela,* Caracas: Tipografía Vargas, 1960.

Hufrauer, G. C. and others. *U. S. Taxation of American Business Abroad, Washington and* Stanford: American Enterprise Institute and Hoover Institution, 1975.

Hussey, Roland D. *The Caracas Company, 1728–1784,* Cambridge: Harvard University Press, 1934.

Inter-American Development Bank. *Venezuela, 1950–1967: Variables, Parameters, and Methodology of the National Accounts,* 2nd edition, Washington: 1968.

International Bank for Reconstruction and Development. *The Economic Development of Venezuela,* Baltimore: Johns Hopkins Press, 1961.

Izard, Miguel. *Series Estadísticas de la Historia de Venezuela,* Mérida: Universidad de Los Andes, 1970.

Jankers, Alfred P. and Malloy, N.M. *Venezuela: Land of Opportunity,* New York: Pageant Press, 1956.

Kelley, R. Lynn. "The 1966 Venezuelan Tax Reform," *Inter-American Economic Affairs,* Vol. 24, Summer 1970, pp. 77–92.

Levine, Daniel H. *Conflict and Political Change in Venezuela,* Princeton: Princeton University Press, 1973.

Levy, Fred D., Jr. *Economic Planning in Venezuela,* New York: Praeger, 1968.

Lieuwin, Edwin. *Petroleum in Venezuela: A History,* Berkeley: University of California Press, 1954.

———. *Venezuela,* 2nd edition, London: Oxford University Press, 1965.

Lombardi, John V. *The Decline and Abolition of Negro Slavery in Venezuela, 1820–1854,* Greenwood Press, Westport, Ct.: 1971.

López, José Eliseo. *La Expansión Demográfica de Venezuela,* Mérida: Universidad de Los Andes, 1968.

———. *Tendencias Recientes in la Población Venezolana,* Mérida: Universidad de Los Andes, 1968.

Losada Andana, Ramón. *Venezuela: Latifundio y Subdesarrollo,* Caracas: Universidad Central de Venezuela, 1969.

Lott, Leo B. *Venezuela and Paraguay: Political Modernity and Tradition in Conflict,* New York: Holt, Rinehart and Winston, 1972.

Luzardo, Rodolfo. *Notas Histórico-económicas, 1928–1963,* Caracas: Editorial Sucre, 1963.

———. *Venezuela: Business and Finance,* Englewood Cliffs: Prentice Hall, 1957.

Machado Gómez, Alfredo. *Crisis y Recuperación: La Economía Monetaria Venezolana entre 1961 y 1968,* Caracas: Banco Central de Venezuela, 1972.

Malavé, Héctor. *Petróleo y Desarrollo Económico de Venezuela,* Caracas: Pensamiento Vivo, 1962.

———. *Formación Histórico del Antidesarrollo de Venezuela,* Caracas: Fondo Editorial Salvador de la Plaza, 1974.

Mancera Galletti, Angel. *Venezuela: Historia de Subdesarrollo,* Caracas: Empresa El Cojo, 1963.

Marrero, Luis. *Venzuela y sus Recursos,* Caracas: Ediciones Cultural Venezolana, 1964.

Marsland, William D. and Amy L. *Venezuela Through Its History,* New York: Thomas Y. Crowell, 1954.

Martínez, Aníbal R. *Cronología del Petróleo Venezolana,* Caracas: Editorial Librería Historia, 1970.

———. *Historia Petrolera Venezolana en 20 Jornadas,* Caracas: Edreca, 1973.

———. *Nuestro Petróleo,* Madrid: Gráficas Minerva, 1963.

———. *Our Gift, Our Oil,* Vienna: Reidel-Dorbrecht, 1966.

———. *Petróleo: Seis Ensayos,* Caracas: Edreca Editores, 1971.

Martz, John D. *Acción Democrática: Evolution of a Modern Political Party in Venezuela,* Princeton: Princeton University Press, 1966.

———. "The Growth and Democraticization of the Venezuelan Labor Movement," *Inter-American Economic Affairs,* Vol. 17, August 1963, pp. 3–18.

Masur, Gerland. *Simón Bolívar,* Albuquerque: University of New Mexico Press, 1948.

Mayobre, José Antonio. *La Verdad sobre Nuestro Petróleo?,* Caracas: Oficina Central de Información, 1967.

Maza Zavala, Domingo F. *Aspectos de Desarrollo Económico de Venezuela,* Caracas: Universidad Central de Venezuela, 1962.

———. *La Economía Venezolana,* Caracas: Ediciones Mundo Económico, 1961.

———. *Hacia la Independencia Económica,* Caracas: Editora de Publicaciones, 1960.

———. *Problemas de Economía Exterior de Venezuela,* Caracas; Universidad Central de Venezuela, 1962.

———. *Venezuela: Una Economíca Dependiente,* Caracas; Universidad Central de Venezuela, 1964.

Maza Zavala, Domingo F. and others. *Venezuela: Crecimiento sin Desarrollo,* Caracas; Editorial Nuestro Tiempo, 1974.

Mejía Alarcón, Pedro. *La Industria del Petróleo en Venezuela,* Caracas: Universidad Central de Venezuela, 1972.

———. *Monopolo y Precios del Petróleo,* Caracas: Universidad Central de Venezuela, 1964.

Mieres, Francisco. *El Petróleo y la Problemática Estructural Venezolana,* Caracas: Universidad Central de Venezuela, 1969.

Mikesell, Raymond F. *Foreign Investment in the Petroleum and Mineral Industries,* Resources for the Future, Baltimore: Johns Hopkins Press, 1971.

Moll, Roberto. *Lecciones de Economía Venezolana,* Caracas: Litografía del Comercio, 1944.

Montiel Ortega, Leonardo. *Nacionalismo e Industrialización,* Caracas: Editorial Pensamiento Vivo, 1962.

———. *Política de Ajustes al Desarrollo Industrial Venezolano,* Caracas: Ministerio de Fomento, 1967.

Morales Crespo, Eddie. *Desarrollo e Industrialización,* Caracas: Corporación Venezolana de Fomento, 1965.

———. *Desarrollo Industrial* Oficina Central de Información, Caracas: 1965.

Morón, Guillermo. *A History of Venezuela,* London: Allen and Unwin, 1964.

Oficina Central de Coordinación y Planificación. *Católogo de Series Estadísticas,* Caracas: CORDIPLAN, 1965.

———. *La Planificacón en Venezuela,* Caracas: CORDIPLAN, 1973.

———. *Plan de la Nación, 1960–1963; Plan de la Nación, 1963–1966; Plan de la Nación, 1965–1968; IV Plan de la Nación, 1970–1974,* Caracas: CORDIPLAN, various dates.

Organization of American States. *El Esfuerzo Interno y las Necesidades al Financimiento Externo para el Desarrollo de Venezuela,* Washington: OAS, 1972.

Orta, Celio S. "Ensayo acerca del Desarrollo Agrícola Venezolano durante el Período 1950–1969," *Revista de Economía y Ciencias Sociales,* Año XII, No. 4, October–December 1970, pp. 5–52.

Pan American Union, *A Statement of the Laws of Venezuela in Matters Affecting Business,* 3rd edition, Washington: Pan American Union, 1962.

———. *Constitution of the Republic of Venezuela, 1961,* Washington: Pan American Union, 1961.

———. *Desarrollo Económico y Social de Venezuela,* Washington: Pan American Union, 1962.

Peattie, Lisa Redfield. *The View from the Barrio,* Ann Arbor: University of Michigan Press, 1968.

Peltzer, Ernesto. *Ensayos sobre Economía,* Caracas: Banco Central de Venezuela, 1965.

Peñaloza, Humberto. *El Cuadro Político del Petróleo Venezolano,* Caracas: Publicaciones Mito Juan, 1970.

Perales, P. *Manual de Geografía Económica de Venezuela,* Caracas: Ediciones Jaime Villegas, 1955.

Pérez Alfonzo, Juan Pablo, *La Dinámica del Petróleo en el Progreso de Venezuela,* Universidad Central de Venezuela, 1965.

———. *El Pentágono Petrolero,* Caracas: Ediciones Revista Política, 1967.

———. *Petróleo, Jugo de la Tierra,* Caracas: Editorial Arte, 1960.

———. *Petróleo y Dependencia,* Caracas: Síntesis Dosmil, 1971.

———. *Venezuela y su Petróleo, Lineamientos de una Política,* Caracas: Imprenta Nacional, 1960.

Pérez Jiménez, Marcos. *Frente a la Infamia.* Caracas: Ediciones Garrido, 1968.

Perfiles de la Economía Venezolana, Caracas: Universidad Central de Venezuela, 1964.

Pernaut, Manuel. *Diez Años de Desarrollo Económico y Social en Venezuela,* Caracas: Ediciones del Cuatricentenario de Caracas, 1966.

Picón Salas, Mariano, y colegas. *Venezuela Independiente 1810–1960,* Caracas: Fundación Eugenio Mendoza, 1962.

Pinto, Gustavo. *Agricultura y Desarrollo: El Caso Venezolano,* Maracaibo: Sociedad Venezolana de Ingenieros Agrónomos, 1966.

Plaza, Salvador de la. *El Petróleo en la Vida Venezolana,* Caracas: Pensamiento Vivo, 1962.

———. *Venezuela, País Privilegiado?* Caracas: Universidad Central de Venezuela, 1973.

Plaza, Salvador de la; and Risquez Iribarren, William; and Guenere Añez, Victor. *Breve Historia del Petróleo y su Legislación en Venezuela,* Caracas: Grafiúnica, 1973.

Pogue, Joseph E. *Oil in Venezuela: An Economic Study,* New York: Chase National Bank, 1949.

Powell, John Duncan. *Political Mobilization of the Venezuelan Peasant,* Cambridge: Harvard University Press, 1971.

Quintero, Rudolfo. *La Cultura de Petróleo,* Caracas: Universidad Central de Venezuela, 1968.

Rangel, Domingo Alberto. *Los Andinos en el Poder,* Mérida: Universidad de Los Andes, 1964.

———. *Capital y Desarrollo: La Venezuela Agraria,* Vol. I; and *El Rey Petróleo*, Vol. II, Caracas: Universidad Central de Venezuela, 1969. *La Oligarquía de Dinero,* Vol. III, Caracas: Editorial Fuentes, 1972.

———. *Gómez, el Amo del Poder,* Valencia: Vadell Hermanos, 1975.

———. *La Industrialización de Venezuela,* Caracas: Pensamiento Vivo, 1958.

———. *El Proceso del Capitalismo Contemporáneo en Venezuela,* Caracas: Universidad Central de Venezuela, 1968.

———. *La Revolución de las Fantasías,* Caracas: Ediciones OFIDI, 1966.

———. *Venezuela, País Ocupado,* Edición Pensamiento Vivo, Caracas, 1960.

Roy, Talton F. *The Politics of the Barrios of Venezuela,* Berkeley: University of California Press, 1969.

Rodwin, Lloyd, and associates. *Planning Urban Growth and Regional Development: The Experience of the Guayana Project of Venezuela,* Cambridge: MIT Press, 1969.

Rourke, Thomas. *Gómez, Tyrant of the Andes,* New York: Morrow, 1936.

Royal Institute of International Affairs. *Venezuela: A Brief Political and Economic Survey,* London: Oxford University Press, 1958.

Salcedo Bastardo, J. L. *Historia Fundamental de Venezuela,* Caracas: Universidad Central de Venezuela, 1970.

Sampson, Anthony. *The Seven Sisters,* New York: Viking Press, 1975.

Shaffer, Edward H. *The Oil Import Program of the United States,* New York: Praeger, 1968.

Shoup, Carl S., and others. *The Fiscal System of Venezuela,* Baltimore: Johns Hopkins Press, 1959.

Silva Guillen, Rafael. *La Reforma Agraria en Venezuela,* Caracas: Instituto Agraria Nacional, 1962.

Tamoí, Ladislao T. *El Nuevo Ideal Nacional de Venezuela,* Madrid: Edciones Verdad, 1954.

Taylor, Phillip B., Jr. *The Venezuela Golpe de Estado de 1958: The Fall of Marcos Pérez,* Washington: Institute for the Comparative Study of Political Systems, 1968.

———. *Venezuela, 1969,* Office of International Affairs, Houston: University of Houston, 1971.

Taylor, Wayne C. and Lindemand, John. *United States Business Performance Abroad: Case Study of the Creole Petroleum Corporation in Venezuela,* Washington: National Planning Association, 1955.

Tejero, José M. and Gómez, Henry. *La Banca Comercial en Venezuela,* Caracas: Universidad Católica de Andŕes Bello, 1967.

Tovar, Ramón. *Imágen Geoeconómico de Venezuela,* Caracas: Viloria y Cruz Editores, 1969.

———. *La Población de Venezuela,* Caracas: Universidad Central de Venezuela, 1968.

Tinoco, Pedro R., hijo. *Petróleo, Factor de Desarrollo,* Caracas: Gráficas Edición de Arte, 1968.

Torrealba Alvarez, Raúl. *El Petróleo en la Economía Venezolana,* Caracas: Editorial 125, 1974.

Tugwell, Franklin. *The Politics of Oil in Venezuela,* Stanford: Stanford University Press, 1975.

Uslar Pietri, Arturo. *Da una a otra Venezuela,* Caracas: Ediciones Mesa Redonda, 1950.

———. *Sumario de Economía Venezolana,* 2a edición, Caracas: Fundación Mendoza, 1958.

Vallenilla, Luis, *Auge, Declinación y Porvenir del Petróleo Venezolano,* Caracas: Editorial Nuevo Tiempo, 1973.

———. *Oil: The Making of a New Economic Order,* New York: McGraw Hill, 1975.

Vallenilla Lanz, Laureano. *Cesarismo Democrático,* 4a edición, Caracas: Tipografía Garrida, 1961.

Velásquez, Ramón. *La Caída del Liberalismo Amarillo,* Caracas: Cromotip, 1973.

Venegas Filardo, Pascual. *Aspectos Geoeconómicos de Venezuela,* Caracas: Ministerio de Relaciones Interiores, 1958.

———. *Siete Ensayos sobre Economía de Venezuela,* Caracas: Monte Avila, 1970.

Vila, Marco Aurelio. *Geografía de Venezuela,* Caracas: Fundación Mendoza, 1961.

Weil, Thomas, and others. *Area Handbook for Venezuela,* Washington: American University, 1971.

Wilgus, A. Curtis, editor. *The Caribbean: Venezuelan Development, a Case History,* Gainesville: University of Florida, 1963.

Wohlrabe, Raymond A. *The Land and People of Venezuela,* Philadelphia: Lippincott, 1959.

TABLES

Table 1. Venezuelan National Income and Its Components, 1950–1969

	National Income (million Bs)	Growth of National Income (%)	National Income per Capita (Bs)	Growth of Nat'l Income per Capita (%)	Payment to Labor (million Bs)	Labor Pay't per worker, including unemployed (Bs)	Labor Payment per employed worker (Bs)	Share of Labor in Nat'l Income (%)	Payment to Capital (million Bs)	Capital Payment Per Unit of Capital stock (%)	Share of Capital in [illegible]
1950	8,607	—	1,702	—	5,147	3,005	3,205	59.8	3,460	13.0	40
1951	9,547	10.9	1,801	5.8	5,494	3,087	3,322	57.5	4,053	14.2	42
1952	10,236	7.2	1,847	2.6	5,664	3,068	3,278	55.3	4,572	14.6	44
1953	10,093	6.5	1,884	2.0	6,317	3,306	3,506	57.9	4,586	13.3	42
1954	12,154	11.5	2.016	7.0	6.874	3.479	3,680	56.6	5,280	13.9	43
1955	13,057	7.4	2,082	3.3	7,280	3,574	3,790	55.8	5,777	14.1	44
1956	14,712	12.7	2,258	8.5	8,129	3,875	4,150	55.3	6,583	14.9	44
1957	16,782	14.1	2,484	10.0	8,786	4,071	4,408	52.4	7,996	16.7	47
1958	18,279	8.9	2.611	5.1	9,958	4,494	4,939	54.5	8,321	16.5	45
1959	19,742	8.0	2,726	4.4	11,422	5,025	5,635	57.8	8,320	15.5	42
1960	19,372	–1.9	2,588	–5.1	11,645	5,002	5,734	60.1	7,727	14.0	39
1961	19,988	3.2	2,532	–0.2	12,151	5,078	5,921	60.8	7,837	14.0	39
1962	21,517	7.6	2,688	4.1	12,469	5,071	5,943	57.9	9,048	15.8	42
1963	23,563	9.5	2,844	5.8	13,780	5,451	6,374	58.5	9,783	16.9	41
1964	27,595	17.1	3,219	13.2	15,326	5,890	6,594	55.5	12,269	20.4	44
1965	29,467	6.8	3,320	3.1	16,284	6,074	6,663	55.3	13,183	21.1	44
1966	30,669	4.1	3,338	0.5	17,452	6,314	6,936	56.9	13,217	20.3	43
1967	32,508	6.0	3,416	2.3	18,827	6,601	7,219	57.9	13,681	20.3	42
1968	34,953	7.0	3,545	3.8	20,099	6,818	7,212	57.5	14,854	21.0	42
1969	36,615	4.8	3,584	1.1	21,244	6,895	7,384	58.0	15,371	20.8	42

Source: *La Economía Venezolana en los Ultimos Treinta Años,* Banco Central de Venezuela, Caracas, 1971.

Table 2. Venezuelan Gross Territorial Product, 1950–1969

	National Income (million Bs)	Depreciation (million Bs)	External Factor Payments (million Bs)	Gross Terr. Product at Factor Costs (million Bs)	Indirect Taxes less Subsidies (million Bs)	Gross Terr. product - Market Prices (million Bs)	Nat'l Income as % of GTP	GTP Per Capita (Bs)	GTP in 1957 Bs	Rate of Growth—GTP (1957 Bs)	Implicit Price Changes (Index)	GTP Per Capita (1957 Bs)	Ratio of External to Internal Factor Payment (%)
1950	8,607	837	1,269	10,713	1,113	11,826	72.8	2,338	12,727	—	92.0	2,516	14.7
1951	9,547	984	1,379	11,910	1,097	13,007	73.4	2,454	14,212	11,7	91.5	2,681	14.4
1952	10,236	1,142	1,454	12,832	1,149	13,981	73.2	2,522	15,247	7.3	91.7	2,751	14.2
1953	10,903	1,259	1,458	13,620	1,186	14,806	73.6	2,559	16,190	6.2	91.5	2,798	13.4
1954	12,154	1,287	1,606	15,047	1,330	16,377	74.2	2,716	17,749	9.6	91.7	2,944	13.2
1955	13,057	1,464	1,906	16,427	1,466	17,893	73.0	2,853	19,325	8.9	92.6	3,081	14.6
1956	14,712	1,693	2,470	18,875	1,525	20,400	72.1	3,131	21,366	10.6	95.5	3,280	16.8
1957	16,782	2,014	3,251	22,047	1,800	23,847	70.4	3,529	23,847	11.6	100.0	3,529	19.4
1958	18,279	2,155	2,097	22,531	2,054	24,585	74.3	3,512	24,164	1.3	101.7	3,452	11.5
1959	19,742	2,370	1,728	23,840	2,117	25,957	76.1	3,584	26,065	7.9	98.1	3,599	8.8
1960	19,372	2,422	2,097	23,891	1,780	25,671	75.5	3,429	27,116	4.0	94.7	3,622	10.8
1961	19,988	2,544	2,322	24,854	2,143	26,997	74.0	3,487	28,479	5.0	94.8	3,679	11.6
1962	21,517	2,855	2,725	27,097	2,428	29,525	72.9	3,688	31,082	9.1	95.0	3,882	12.7
1963	23,563	2,922	2,853	29,338	2,848	32,186	73.2	3,885	33,222	6.9	96.9	4,010	12.1
1964	27,595	3,051	3,223	33,869	1,768	35,637	77.4	4,157	36,448	9.7	97.8	4,251	11.7
1965	29,467	3,211	3,492	36,170	1,755	37,925	77.7	4,273	38,605	5.9	98.2	4,350	11.9
1966	30,669	3,496	3,393	37,558	1,958	39,516	77.6	4,300	39,492	2.3	100.1	4,298	11.1
1967	32,508	3,848	3,272	39,628	1,995	41,625	78.1	4,374	41,085	4.0	101.3	4,317	10.1
1968	34,953	4,229	3,532	42,714	2,134	44,848	77.9	4,549	42,268	5.3	106.1	4,287	10.1
1969	36,615	4,538	3,488	44,641	2,226	46,867	78.1	4,588	44,787	3.5	104.6	4,384	9.5

Source: *La Economía Venezolana en los Ultimos Treinta Años,* Banco Central de Venezuela, Carcas, 1971.

Table 3. Venezuelan Gross Territorial Product by Sectors, 1950–1969 (million 1957 bolívares)

	Agriculture	Petroleum	Mining	Manufacturing	Construction	Water and Electricity	Transportation and and Communication	Commerce	Services	Tertiary Sector	Primary Sector	Secondary Sector	Gross Territorial Product
1950	1,014	3,797	20	1,274	827	69	699	1,726	3,301	5,726	4,831	2,170	12,727
1951	1,156	4,323	44	1,357	1,032	83	706	1,983	3,528	6,217	5,523	2,472	14,212
1952	1,239	4,594	75	1,578	1,144	97	695	2,085	3,760	6,520	5,908	2,819	15,247
1953	1,280	4,500	79	1,772	1,220	119	843	2,291	4,086	7,220	5,859	3,111	16,190
1954	1,284	4,842	152	2,022	1,376	135	876	2,678	4,384	7,935	6,278	3,533	17,749
1955	1,352	5,514	221	2,267	1,363	159	951	2,862	4,636	8,449	7,087	3,789	19,325
1956	1,446	6,300	279	2,484	1,605	187	946	3,156	4,963	9,065	8,025	4,276	21,366
1957	1,507	7,135	383	2,766	1,581	238	940	3,933	5,364	10,237	9,025	4,585	23,847
1958	1,576	6,708	379	2,972	1,618	281	1,002	3,803	5,825	10,630	8,663	4,871	24,164
1959	1,642	7,154	420	3,418	1,707	336	1,090	4,003	6,295	11,388	9,216	5,461	26,056
1960	1,987	7,325	463	3,347	1,647	371	1,001	3,976	6,989	11,976	9,775	5,365	27,116
1961	1,999	7,483	358	3,548	1,471	422	996	3,811	8,381	13,188	9,850	5,411	28,470
1962	2,103	8,214	333	3,869	1,420	501	996	4,051	9,595	14,642	10,650	5,790	31,082
1963	2,220	8,336	282	4,148	1,340	590	1,045	4,315	10,946	16,306	10,838	6,078	33,222
1964	2,398	8,734	390	4,667	1,496	663	1,181	4,932	11,987	18,100	11,522	6,826	36,448
1965	2,546	8,915	425	5,049	1,545	738	1,267	5,455	12,665	19,387	11,886	7,332	38,605
1966	2,653	8,653	427	5,100	1,648	808	1,306	5,695	13,202	20,203	11,733	7,556	39,492
1967	2,798	9,093	411	5,375	1,682	894	1,370	5,907	13,555	20,832	12,302	7,951	41,085
1968	2,938	9,278	366	5,682	2,040	1,041	1,460	6,344	14,119	21,923	12,582	8,763	43,268
1969	3,030	9,222	454	5,971	1,975	1,147	1,466	6,501	15,075	23,042	12,706	9,039	44,787

Source: *La Economía Venezolana en los Ultimos Treinta Años,* Banco Central de Venezuela, Caracas, 1971.

Table 4. Venezuelan Gross Territorial Product and National Income, 1970–1974
(millions of bolívares)

	Gross Territorial Product	Indirect Taxes less Subsidies	Depreciation	Domestic Factor Income	Labor Payments	Other Payments	Payments by Petroleum to Government	National Income	National Income - Market Prices
1970	51,991	2,848	4,794	44,349	20,923	23,426	6,200	41,308	44,156
1971	57,005	2,730	5,263	49,012	23,011	26,001	7,546	45,075	47,805
1972	63,305	2,669	5,397	55,239	25,491	29,748	8,411	50,599	53,268
1973	76,341	2,824	5,724	67,793	28,583	39,230	13,389	61,073	63,897
1974	126,699	2,926	6,190	117,583	35,925	81,656	38,558	99,334	102,260

Source: *Informe Económico, 1974,* Banco Central de Venezuela

Table 5. Venezuelan Gross Territorial Product by Sector, 1970–1974
(millions of 1968 bolívares)

	Agriculture	Crude Petroleum	Mining	Industry	Electricity	Construction
1970	3,490	9,195	687	8,018	859	1,902
1971	3,458	8,607	650	8,582	949	2,222
1972	3,517	7,960	618	8,890	1,012	2,783
1973	3,747	8,531	807	9,152	1,122	3,022
1974	4,030	7,524	935	9,639	1,256	2,809

Commerce	Transport & Commu nication	Financial & Business Services	Community Services	Government	Other	Gross Territorial Product
5,416	4,857	6,093	2,067	5,278	2,210	50,072
5,494	5,230	6,345	2,367	5,540	2,022	51,466
5,695	5,735	6,719	2,557	5,945	1,897	53,328
5,957	5,950	7,135	2,868	6,198	1,942	56,431
6,446	6,657	7,498	3,087	7,128	1,980	58,989

Source: *Informe Económico, 1974,* Banco Central de Venezuela

Table 6. Venezuelan Income and Taxes of Central Government, 1950–1969

	Total Income	Petroleum Income Tax	Other Income Tax	Customers and Consuler Duties	Cigarette and Liquor Taxes	Other In-direct Taxes*	Petroleum Royalties	Other Royalties	Borrowing	Other Income**
1950	1,917	273	101	391	132	278	614	118	0	10
1951	2,267	394	110	419	154	192	864	130	0	4
1952	2,408	525	123	452	165	203	822	109	0	9
1953	2,534	594	144	452	168	226	826	122	0	2
1954	2,632	507	169	485	163	268	901	139	0	0
1955	2,992	585	199	515	156	327	1,034	154	0	22
1956	4,380	712	233	535	196	333	1,211	177	0	983
1957	5,405	931	277	628	228	441	1,539	219	0	1,142
1958	4,706	1,198	400	653	262	571	1,362	259	0	1
1959	5,743	1,465	520	688	300	529	1,638	300	302	1
1960	6,147	1,260	545	514	291	373	1,632	352	1,180	0
1961	7,074	1,555	785	435	350	841	1,574	253	1,224	57
1962	6,589	1,500	673	414	360	1,123	1,602	239	676	4
1963	6,619	1,759	726	404	379	1,372	1,715	241	23	0
1964	7,217	2,155	804	478	420	164	2,499	266	84	347
1965	7,367	2,188	1,029	443	452	219	2,531	312	102	91
1966	7,952	2,355	1,294	480	461	282	2,557	322	101	100
1967	8,678	2,929	1,236	479	516	253	2,737	388	56	84
1968	9,159	3,046	1,292	510	541	270	2,744	372	244	140
1969	9,680	2,694	1,417	552	563	281	2,749	408	870	146

* Includes profits from exchange trading
** Includes petroleum concessions, coinage, revaluation of exchange rate
Source: *La Economía Venezolana en los Ultimos Treinta Años,* Banco Central de Venezuela, Caracas, 1971

Table 7. Venezuelan Income, Expenditure, Borrowing and Treasury Reserve, 1949–1969
(000,000 bolívares)

	Recurring Income	Ordinary Expenditures	Balance	Special Income*	Special Expenditures	Adjusted Balance	Increase (+) Decrease (−) in Public Debt	Increase (+) Decrease (−) in Treasury Reserve	Treasury Balance (year-end)	Total Expenditures	New Borrowing	Extraordinary Income	Tax and Royalty Income	Total Income
1949	1,976	1,955	21	4	—	25	—	25	276	1,955	—	4	1,977	1,980
1950	1,907	2,074	–167	10	—	–157	—	–157	119	2,074	—	10	1,907	1,917
1951	2,263	2,294	–31	4	—	–27	—	–27	92	2,294	—	4	2,262	2,267
1952	2,399	2,384	–15	9	—	24	—	24	116	2,384	—	9	2,399	2,408
1953	2,532	2,405	127	2	—	129	—	129	245	2,405	—	2	2,532	2,534
1954	2,632	2,592	40	—	—	40	—	40	285	2,592	—	—	2,632	2,632
1955	2,970	2,983	–13	22	—	9	—	9	294	2,983	—	22	2,970	2,992
1956	3,397	3,348	49	983	—	1,032	—	1,032	1,326	3,348	—	983	3,397	4,380
1957	4,263	4,376	–113	1,142	—	1,029	—	1,029	2,355	4,376	—	1,142	4,263	5,405
1958	4,705	6,239	–1,534	1	—	–1,533	–21	–1,554	801	6,260	—	1	4,705	4,706
1959	5,440	6,294	–854	1	—	–853	281	–571	230	6,314	302	1	5,440	5,743
1960	4,967	5,820	–853	1	—	–852	853	1	231	6,147	1,180	—	4,967	6,147
1961	5,185	6,027	–842	665	—	–177	176	–1	230	7,075	1,224	57	5,793	7,074
1962	5,383	5,270	113	529	—	642	–312	331	561	6,258	676	2	5,911	6,589
1963	6,333	5,983	350	263	—	613	–584	29	589	6,590	23	—	6,596	6,619
1964	6,786	6,444	342	347	—	689	–572	117	707	7,100	84	347	6,786	7,217
1965	7,174	7,118	56	91	—	147	–180	–33	674	7,400	102	91	7,174	7,367
1966	7,632	7,660	–28	219	—	191	–163	28	701	7,924	101	100	7,751	7,952
1967	8,069	8,325	–256	554	—	298	–224	74	767**	8,605	56	84	8,539	8,679
1968	8,524	8,821	–297	391	98	–4	–11	–15	759**	9,174	244	140	8,775	9,159
1969	8,643	9,545	–902	167	118	–853	707	–146	613	9,826	870	146	8,664	9,680

* Includes petroleum and mining back taxes, adjustments in collection system, revaluation of exchange rate, coinage and petroleum concessions.

** Does not match change in Treasury reserves because of special adjustement for previous years.

Table 8. Venezuelan General Government Income, 1970–1974 (millions of bolívares)

	1970	1971	1972	1973	1974
Total	10,252	12,123	12,547	16,433	42,799
Income Tax	4,561	6,563	7,062	10,151	29,407
Petroleum Income Tax	2,786	4,743	5,057	7,798	26,818
Mining Income Tax	245	269	173	176	354
Other Income Tax	1,472	1,524	1,799	2,174	2,233
Indirect Taxes	1,542	1,666	1,786	1,885	2,736
Customs Duties	422	428	468	468	1,056
Petroleum Royalties	2,864	2,873	2,794	3,381	9,628
Foreign Borrowing	476	9	231	61	48
Domestic Borrowing	278	477	120	284	101

Source: *Informe Económico, 1974,* Banco Central de Venezuela, 1974.

Table 9. Venezuelan Central Government Expenditures, 1950–1959
(000,000 bolívares)

	Total Expenditures	Current Expenditures	Goods and Non-Personal Services	Personal Services	Total Transfers	Transfers to Autonomous Institutes	Consumer and Producer Subsidies	Transfers to States and Municipalities	Maintenance and Repairs	Capital Expenditures	New Assets and Other Goods	Existing Assets
1950	2,074	1,574	205	621	698	392	39	263	50	500	469	27
1951	2,294	1,674	207	680	710	402	47	270	79	607	619	35
1952	2,363	1,707	222	726	690	353	46	296	70	649	603	40
1953	2,435	1,693	235	749	651	331	50	298	59	712	681	23
1954	2,592	1,837	261	771	746	377	77	290	60	755	722	26
1955	2,983	2,061	308	824	879	487	77	309	51	922	872	45
1956	3,348	2,148	338	805	956	584	48	318	48	1,200	1,119	76
1957	5,348	2,759	409	822	1,390	973	44	366	138	1,617	1,528	85
1958	6,256	4,461	550	1,284	2,261	1,639	83	533	366	1,798	1,702	93
1959	6,314	4,843	552	1,708	2,180	1,488	106	644	400	1,494	1,393	93

Source: *Memoria, 1959,* Banco Central de Venezuela, Caracas, 1959

Table 10. Venezuelan Central Government Expenditures, 1960–1969
(000,000 bolívares)

	Total Expenditures	Current Expenditures	Goods	Personal Services	Non-Personal Services	Maintenance	Transfers	Consumer	Producer	States and Municipalities	Administration Units	Capital Expenditures	New Assets	Existing Assets	Financial Investment	Capital Transfers	State Enterprises	States and Municipalities	Amortization Of Public Debt	Internal Debt	External Debt
1960	6,147	3,937	325	1,797	185	196	1,434	212	26	513	674	2,210	986	21	20	856	718	138	327	310	17
1961	7,075	3,996	254	1,898	187	191	1,466	186	39	516	717	3,079	911	39	77	956	827	129	1,048	937	111
1962	6,258	3,717	292	1,737	201	88	1,399	212	52	572	552	2,541	655	31	80	761	620	141	988	760	228
1963	6,590	4,224	366	2,045	274	72	1,467	150	64	707	533	2,366	852	50	42	773	612	161	607	383	224
1964	7,100	4,329	262	2,216	198	89	1,564	171	34	749	594	2,771	945	51	55	1,042	866	176	656	493	163
1965	7,400	4,849	339	2,434	221	52	1,803	201	49	819	717	2,551	1,025	11	60	1,163	973	190	282	208	74
1966	7,924	5,137	365	2,675	220	50	1,827	177	45	933	652	2,787	1,006	31	115	1,333	1,110	223	264	185	79
1967	8,605	5,412	458	2,664	229	44	2,017	207	38	1,011	745	3,193	1,185	31	200	1,477	1,243	234	280	184	96
1968	9,174	5,782	444	2,862	278	51	2,147	225	44	1,049	813	3,392	1,331	29	119	1,642	1,343	281	255	162	93
1969	9,826	6,030	410	2,956	314	64	2,286	208	31	1,094	933	3,796	1,780	9	94	1,727	1,431	256	163	136	27

Source: *Informe Económico, 1969,* Banco Central de Venezuela, Table A-IX-3.

Table 11. Venezuelan Expenditures by Functions, 1960–1969 (000,000 bolívares)

	1960	1961	1962	1963	1964	1965	1966	1967	1968	1969
Political and Administrative	1,040	1,028	900	1,144	1,163	1,263	1,378	1,532	1,638	1,545
General Administration	228	225	197	256	263	262	296	342	365	360
Defense	588	581	509	613	650	741	782	876	943	844
Police and Justice	184	182	159	181	207	209	236	243	254	261
Exterior Relations	40	40	35	35	41	50	62	67	72	76
Electoral Council	—	—	—	59	2	1	2	4	4	4
Social	1,605	1,733	1,631	1,741	2,078	2,422	2,631	3,046	3,304	3,905
Education	572	636	627	687	811	926	930	1,182	1,245	1,467
Sanitation	375	447	451	489	536	615	652	795	837	864
Social Assistance	250	247	210	232	262	268	378	183	192	336
Water Supply	147	146	124	174	194	211	274	281	296	371
Recreation and Entertainment	12	12	10	9	14	19	31	25	26	43
Housing and Urbanism	226	223	190	130	240	365	345	560	687	801
Employer-Employee Relations	23	22	19	20	21	18	21	20	21	23
Economic	2,468	2,429	1,824	2,098	2,199	2,303	2,354	2,400	2,536	2,715
Agriculture	534	660	545	569	702	791	775	849	897	906
Electricity	256	171	118	139	182	87	121	83	88	122
Industry	375	510	221	154	170	318	331	362	382	147
Transport and Storage	901	820	755	996	882	817	854	848	896	1,170
Communications	298	198	137	162	165	188	182	156	165	173
Extractive Industry	87	58	40	63	88	89	80	93	99	112
Tourism	17	12	8	15	10	13	11	9	9	85
Other	1,034	1,885	1,903	1,607	1,660	1,412	1,561	1,627	1,696	1,661
Interest on Public Debt	25	38	40	60	58	61	56	57	61	86
Debt Amortization	327	1,048	988	607	656	282	264	280	255	163
Administrative Debt	—	48	—	—	—	15	41	20	16	23
Transfers to Regional Govts.	655	667	757	870	934	1,040	1,171	1,245	1,330	1,350
Acquisition of Financial Assets	20	77	111	65	6	9	23	19	27	32
Roman Catholic Church	7	7	7	5	6	5	6	6	7	7
Total	6,147	7,075	6,258	6,590	7,100	7,400	7,924	8,605	9,194	9,826

Source: *Informe Económico, 1969,* Banco Central de Venezuela, Caracas, Table A-IX-10.

Table 12. Expenditures of Central Government by Category and Function; 1970–1974
(millions of bolívares)

	1970	1971	1972	1973	1974
Current Expenditures	7,159	8,099	8,968	10,320	15,604
Goods and Services	4,603	5,372	5,747	6,341	8,239
Transfers	2,556	2,727	3,221	3,979	7,365
Investment Expenditures	2,959	3,099	3,445	4,252	23,867
Capital formation	1,298	1,248	1,547	1,696	2,196
Financial Investment	1,496	1,596	1,507	2,076	21,032
Transfers	165	255	391	480	639
Public Debt	177	717	429	470	588
Total	10,295	11,915	12,842	15,042	40,059
Services	823	923	991	1,135	1,668
General Administration	456	554	541	664	819
Foreign Affairs	75	88	92	109	387
Public Order and Security	258	239	306	297	378
General Research	34	42	52	65	84
Defense	898	1,161	1,306	1,363	1,944
Social	3,314	3,607	4,049	5,015	7,339
Education	1,632	1,849	2,237	2,712	3,783
Sanitation	767	795	904	1,056	1,283
Social Security	426	453	432	593	643
Housing and Community Services	423	447	404	555	1,467
Other Community Social Services	66	63	72	99	163

Table 12. Expendtures of Central Government by Category and Function; 1970–1974 (millions of bolívares) (*continued*)

Economic	3,452	3,712	3,878	4,803	12,293
Administration, Regulation, and Research	128	132	87	100	108
Agriculture, Livestock, and Fishing	1,088	1,233	1,149	1,640	4,381
Mining, Manufacturing, and Construction	656	738	812	1,027	4,241
Electricity, Gas and Water	325	309	300	435	1,234
Roads	791	805	1,010	969	1,381
Coastal and Internal Waterways	148	145	159	175	225
Other Transport and Communication	313	334	344	425	636
Other Services	3	16	17	32	87
Other	1,808	2,512	2,618	2,726	16,815
Interest on Public Debt	166	189	164	190	206
Repayment of Public Debt	146	692	411	433	537
Administrative Debt	31	25	18	37	51
Transfers to Regional Government	1,439	1,561	2,000	2,059	2,913
Acquisition of Financial Assets	26	45	25	7	58
Venezuelan Investment Fund	—	—	—	—	13,000
Export Finance Fund	—	—	—	—	—

Source: *Informe Economíco, 1974,* Banco Central de Venezuela, A-200–201

Table 13. Venezuelan Expenditures by Ministry, 1950–1974 (millions of bolívares)

	Total Expenditures	Interior Relations	Justice	Exterior Relations	Treasury	Defense	Development	Mines and Hydrocarbons	Public Works	Education	Health and Social Assistance	Agriculture and Livestock	Labor	Communications
1950	2,074	362		18	129	187	138		739	128	135	118	16	102
1951	2,294	328	68	22	140	195	152	15	836	139	144	110	21	125
1952	2,384	366	75	23	209	208	118	27	798	137	150	129	21	122
1953	2,405	367	76	25	235	231	126	20	744	144	157	124	22	134
1954	2,592	369	78	33	285	239	113	35	807	157	167	126	27	157
1955	2,927	389	80	25	358	266	153	35	965	160	172	148	28	148
1956	3,348	407	81	33	325	286	252	165	1,087	164	190	140	28	190
1957	4,376	421	86	31	463	375	402	128	1,568	173	216	229	33	250
1958	6,259	876	112	36	346	496	597	288	2,002	260	336	465	54	393
1959	6,414	874	160	45	464	669	404	167	2,025	394	407	383	72	349
1960	6,157	812	163	42	722	526	397	98	1,561	506	453	468	58	350
1961	7,076	1,029	146	42	1,326	691	401	114	1,373	563	453	606	54	279
1962	6,258	1,081	139	35	1,347	563	136	102	1,119	576	439	416	50	255
1963	6,590	1,287	156	36	1,000	658	168	87	1,337	674	485	372	51	279
1964	7,100	1,304	173	41	1,018	663	168	110	1,509	745	554	486	61	268
1965	7,400	1,442	184	51	650	748	221	117	1,535	878	638	584	59	294
1966	7,924	1,629	206	62	647	850	195	135	1,680	966	674	528	61	291
1967	8,605	1,733	218	68	724	904	183	159	1,800	1,096	734	621	62	304
1968	9,051	1,817	226	68	749	909	216	180	1,841	1,190	756	662	116	319
1969	9,826	1,773	230	76	745	897	174	184	2,517	1,350	804	661	99	316
1970	10,295	1,867	235	84	724	923	266	201	2,220	1,654	877	767	154	323
1971	11,915	2,039	243	90	1,581	1,171	221	286	2,166	1,874	924	838	146	336
1972	12,842	2,444	276	98	1,526	1,338	225	412	2,097	2,250	1,120	736	85	335
1973	15,042	2,672	278	115	1,754	1,385	284	416	2,540	2,805	1,005	1,207	207	374
1974	40,059	4,352	366	390	16,214	2,022	1,545	1,081	4,515	3,954	1,426	3,377	208	609

Source: *Anuario Estadístico, 1970,* Dirección General de Estadística y Censos Nacionales, and *Informe Económico 1974,* Banco Central de Venezuela.

Table 14. Venezuelan Debt, 1949–1969
(000,000 bolivares)

	Public Debt Total	External Direct Debt	External Indirect Debt	Internal Direct Debt	Internal Indirect Debt	Direct Floating Debt	Indirect Floating Debt
1949	23	0	—	23	—	—	—
1950	15	0	—	15	—	—	—
1951	15	0	—	15	—	—	—
1952	177	0	—	7	170	—	—
1953	305	0	—	45	260	—	—
1954	458	0	—	35	423	—	—
1955	702	0	—	31	671	—	—
1956	699	0	—	27	672	—	—
1857	938	0	—	23	915	—	—
1958	1,138	54	596	19	469	—	—
1959	2,729	37	403	323	501	524	941
1960	2,579	688	298	653	534	187	219
1961	2,778	646	168	1,031	527	187	219
1962	2,577	594	152	874	551	187	219
1963	2,231	427	265	678	645	116	100
1964	1,932	470	386	248	678	93	57
1965	2,132	537	548	204	787	56	—
1966	2,353	607	724	164	839	19	—
1967	2,830	594	953	174	1,109	—	—
1968	3,527	669	1,211	320	1,327	—	—
1969	4,556	1,047	1,309	738	1,462	—	—

Source: *La Economiá Venezolana en los Ultimos Treinta Años,* Banco Central de Venezuela, Caracas, 1971.

Table 15. Venezuelan National Public Debt, 1970–1974
(millions of bolívares)

	1970	1971	1972	1973	1974
DEBT AUTHORIZED BY CONGRESS	5,326	6,308	6,777	7,136	8,651
Direct	2,513	2,408	2,462	2,549	4,526
Indirect	2,813	3,900	4,315	4,587	4,125
EXTERNAL	2,766	3,596	3,919	3,913	3,293
Direct	1,480	1,311	1,301	1,126	879
Indirect	1,286	2,285	2,618	2,787	2,414
INTERNAL	2,560	2,712	2,858	3,223	5,358
Direct	1,033	1,097	1,161	1,423	3,647
Indirect	1,527	1,615	1,697	1,800	1,711
TOTAL DEBT AUTHORIZED BY COUNCIL OF MINISTERS	166	174	433	1,298	1,525
External Indirect	166	174	421	1,288	1,416
Internal Indirect	—	—	12	10	109

Source: *Informe Economíco 1974,* Banco Central de Venezuela, H-223-226

Table 16. Venezuelan Population and Labor Force, 1950–1969
(Year End Estimates, thousands)

	Total Population	Economically Active	Employed	Unemployed	Agriculture	Petroleum	Mining	Manufacturing	Construction	Energy	Transporation & Communication	Commerce	Services
1950	5,058	1,713	1,606	107	709	45	6	188	94	5	55	154	350
1951	5,301	1,780	1,654	126	724	45	5	190	100	6	58	162	364
1952	5,543	1,846	1,728	118	742	46	5	198	109	7	62	174	385
1953	5,786	1,911	1,802	109	756	45	6	208	117	8	67	187	408
1954	6,029	1,975	1,868	107	767	43	9	217	122	9	72	200	429
1955	6,272	2,037	1,921	116	772	43	10	227	124	11	76	211	447
1956	6,515	2,098	1,959	139	770	44	10	234	131	13	79	220	458
1957	6,757	2,158	1,993	165	764	46	11	242	135	14	83	229	469
1958	7,000	2,216	2,016	200	755	44	11	250	143	16	85	236	476
1959	7,243	2,273	2,027	246	741	43	9	253	140	20	88	245	488
1960	7,486	2,328	2,037	297	731	41	10	253	131	22	91	252	499
1961	7,741	2,393	2,052	341	730	37	9	271	113	24	89	239	540
1962	8,006	2,459	2,098	361	722	35	9	299	108	26	85	260	554
1963	8,284	2,528	2,162	366	713	34	9	325	100	29	96	286	570
1964	8,573	2,602	2,324	278	706	33	9	373	115	30	115	345	598
1965	8,875	2,681	2,444	237	701	32	9	409	120	32	128	400	613
1966	9,189	2,764	2,516	248	697	29	9	414	129	33	134	428	643
1967	9,517	2,852	2,608	244	694	27	9	444	134	35	143	455	667
1968	9,857	2,948	2,787	161	692	25	9	471	175	37	159	491	728
1969	10,108	3,081	2,877	204	706	24	9	496	176	31	182	521	732

Source: *La Economía Venezolana en los Ultimos Treinta Años,* Banco Central de Venezuela, Caracas, 1971.

Table 17. Venezuelan Gross Territorial Product per Worker by Sector, 1950–1969 (Bs 000/worker)

	Agriculture	Petroleum	Mining	Manufacturing	Construction	Energy	Transportation & Communication	Commercial	Services	TOTAL
1950	1,430	84,378	3,333	6,777	8,798	13,800	12,709	11,208	9,431	7,876
1951	1,597	96,067	8,800	7,142	10,320	13,833	12,172	12,241	9,692	8,593
1952	1,670	99,870	15,000	7,970	10,495	13,857	10,887	11,983	9,766	8,823
1953	1,693	100,000	13,167	8,519	10,427	14,875	12,582	12,251	10,015	8,984
1954	1,674	112,837	16,889	9,318	11,279	15,000	12,167	13,390	10,219	9,502
1955	1,751	128,233	22,100	9,987	10,992	14,455	12,513	13,564	10,371	10,060
1956	1,878	143,182	27,900	10,615	12,252	14,385	11,975	14,345	10,836	10,907
1957	1,973	155,109	34,818	11,430	11,711	17,000	11,325	17,175	11,437	11,965
1958	2,087	152,455	34,455	11,880	11,315	17,563	11,788	16,114	12,237	11,986
1959	2,216	166,372	46,667	13,510	12,264	16,800	12,386	16,339	12,900	12,859
1960	2,718	178,659	51,444	13,229	12,573	16,864	11,110	15,778	14,066	13,351
1961	2,738	292,514	39,778	13,092	13,018	17,583	11,191	15,946	15,520	13,879
1962	2,913	234,686	37,000	12,940	13,148	19,269	11,718	15,581	17,319	14,815
1963	3,114	245,176	31,333	12,763	13,400	20,345	10,885	15,087	19,204	15,366
1964	3,397	264,667	43,333	12,512	13,009	22,100	10,270	14,296	20,045	15,683
1965	3,632	278,594	47,222	12,345	12,875	23,063	9,898	13,638	20,661	15,796
1966	3,806	298,379	47,444	12,319	12,775	24,485	9,746	13,306	20,532	15,696
1967	4,032	336,778	45,667	12,106	12,552	25,543	9,580	12,982	20,322	15,753
1968	4,256	371,120	40,667	12,064	11,657	28,135	9,182	12,921	19,394	15,525
1969	4,292	384,250	50,444	12,038	11,222	37,000	8,055	12,478	20,594	15,567

Source: *La Economía Venezolana en los Ultimos Treinta Años,* Banco Central de Venezuela, Caracas, 1971

Table 18. Venezuelan Capital Per Worker, 1950–1969 (1957 Bs)

	Total Capital per Worker	Agriculture	Petroleum	Mining	Manufacturing	Construction	Electricity	Transportation and Communication	Commerce	Services
1950	16,515	6,961	62,289	10,833	8,170	2,340	43,400	47,582	12,753	34,763
1951	17,221	6,854	78,911	28,000	8,858	2,290	42,833	50,103	14,198	35,168
1952	18,128	6,973	81,848	48,000	9,263	1,844	44,286	51,532	15,374	36,190
1953	19,058	7,081	95,267	42,000	9,462	1,530	41,875	52,612	16,449	37,662
1954	20,257	7,430	106,116	54,000	9,931	1,811	47,889	59,708	16,245	39,014
1955	21,250	7,756	114,093	70,700	10,154	1,976	44,364	64,868	15,886	40,045
1956	22,477	8,203	128,409	88,700	10,603	2,206	44,077	71,012	15,605	41,035
1957	23,826	8,702	136,891	112,000	11,463	3,911	52,143	68,193	15,996	42,563
1958	25,241	9,139	160,636	110,182	12,392	3,720	77,500	67,176	16,453	44,628
1959	27,072	9,918	173,977	148,889	12,593	3,571	71,000	68,534	16,624	47,125
1960	27,160	9,278	183,732	150,000	15,885	3,382	72,091	73,747	16,139	45,192
1961	27,193	9,701	194,541	162,000	15,923	3,496	60,167	78,225	16,833	42,450
1962	27,317	10,380	195,714	155,889	15,307	3,417	72,885	85,012	15,154	42,522
1963	26,897	11,070	194,265	101,889	14,569	3,750	67,690	78,854	13,483	42,521
1964	25,926	11,830	193,394	100,778	13,445	3,435	68,533	69,783	11,177	42,249
1965	25,662	12,655	194,125	94,000	13,269	3,608	68,750	65,406	9,695	43,194
1966	25,917	13,376	204,241	87,333	14,007	3,403	69,848	65,724	9,061	43,369
1967	25,962	14,122	209,593	81,556	13,685	3,313	68,686	65,573	8,675	43,877
1968	25,493	14,884	228,880	78,444	13,628	2,743	67,270	64,830	8,094	42,103
1969	25,736	15,385	240,042	73,556	13,540	2,807	81,129	60,236	7,708	43,788

Source: *Memoria, 1959, Informe Económica, 1969,* and *La Economía Venezolana en los Ultimos Treinta Años,* Banco Central de Venezuela, Caracas, 1959, 1969, and 1971.

Table 19. Venezuelan Public Money Supply and Liquidity, 1950–1974 (000,000 bolívares)

	Coins	Bills	Demand Deposits	Money Supply	Time & Savings Deposits	Total Liquidity	Annual Variations	Exchange Effect	Fiscal Effect	Credit Effect
1950	99	687	772	1558	203	1760	80	–196	156	120
1951	92	726	805	1623	245	1868	108	–19	27	100
1952	95	801	987	1883	326	2209	341	179	–24	186
1953	97	825	1130	2052	430	2481	272	91	–130	311
1954	100	882	1152	2134	560	2694	213	–64	–40	317
1955	108	925	1312	2345	752	3097	403	110	13	280
1956	100	1012	1567	2679	984	3663	566	1293	–1033	306
1957	105	1181	2223	3509	1554	5061	1398	1563	–1029	864
1958	120	1222	2514	3856	1886	5742	681	–1208	1554	335
1959	120	1255	2261	3636	2134	5770	–52	–1112	571	489
1960	123	1367	1955	3445	1748	5193	–497	–1013	507	9
1961	132	1214	2168	3514	1723	5237	44	–32	72	4
1962	128	1188	2066	3382	1893	5275	38	–19	–580	637
1963	134	1229	2267	3630	2312	5942	667	652	–399	414
1964	142	1284	2689	4115	2720	6834	892	735	–946	1103
1965	178	1386	2968	4532	3011	7542	708	–166	–315	1189
1966	215	1457	2912	4584	2989	7573	31	–249	–291	571
1967	238	1572	3299	5109	3568	8676	1103	496	–321	928
1968	287	1686	3680	5653	4041	9694	1018	149	–139	1008
1969	291	1778	4097	6166	4785	10952	1258	–369	778	849
1970	306	1893	4221	6420	5517	11,937	1032	–279	278	1416
1971	316	2054	5329	7699	6656	14,355	2416	2014	–602	696
1972	329	2288	6021	8638	8224	16,882	2509	644	168	1629
1973	347	2539	7743	10629	10275	20,905	4043	2158	–2404	4333
1974	448	3396	11125	14969	12490	27,459	6554	—	—	—

Source: *La Economía Venezolana en los Ultimos Treinta Años,* Banco Central de Venezuela, Caracas, 1971, for 1950–1969; *Revista del Banco Central de Venezuela* October–December, 1974, for 1970–1974

Table 20. Commercial Bank Assets and Liabilities in Venezuela, 1950–1969
(million of bolívares)

	International Reserves	Loans and Investments	National Balances	Other Accounts	Total Assets and Liabilities	Demand Deposits	Time and Savings Deposits	Total Deposits	Other Liabilities	Captial and Reserves	Number of Banks	Number of Banks, Branches and Agencies
1950	50	841	275	68	1,234	684	157	841	148	245	16	65
1951	49	922	380	73	1,424	814	199	1,013	133	278	19	72
1952	44	1,112	441	99	1,696	985	269	1,254	133	309	19	85
1953	51	1,424	456	125	2,056	1,158	368	1,526	178	352	21	99
1954	55	1,809	446	125	2,435	1,218	484	1,702	292	441	27	130
1955	46	2,125	496	171	2,838	1,432	655	2,087	216	535	27	175
1956	57	2,537	681	240	3,515	1,667	903	2,570	300	645	30	242
1957	47	3,657	1,048	395	5,147	2,431	1,433	3,864	446	837	33	314
1958	90	4,239	1,044	865	6,238	2,857	1,807	4,664	534	1,040	37	384
1959	17	4,896	1,019	587	6,519	2,588	2,033	4,621	750	1,148	37	455
1960	21	4,482	904	361	5,768	2,167	1,667	3,834	765	1,169	37	473
1961	45	4,597	1,148	435	6,225	2,354	1,638	3,992	1,055	1,178	37	469
1962	59	4,646	1,055	417	6,177	2,346	1,804	4,150	833	1,194	37	460
1963	–4	4,869	1,273	613	6,751	2,540	2,196	4,736	800	1,215	36	454
1964	36	5,533	1,226	811	7,606	2,972	2,548	5,520	826	1,260	35	453
1965	72	6,290	1,238	721	8,321	3,344	2,785	6,129	957	1,235	34	450
1966	9	6,388	1,241	931	8,569	3,415	2,739	6,154	1,121	1,294	34	468
1967	21	6,948	1,462	1,001	9,432	3,897	3,239	7,136	968	1,328	34	473
1968	40	7,768	1,661	965	10,434	4,351	3,592	7,943	1,027	1,464	34	479
1969	35	8,400	2,003	1,065	11,503	4,735	4,059	8,794	1,158	1,551	31	506

*Includes deposits in foreign money and government deposits.
Source: *La Economía Venezolana en los Ultimos Treinta Años,* Banco Central de Venezuela, Caracas, 1971

Table 21. Venezuelan Commercial Bank Position, 1970–1974
(Year-end balances in millions of bolívares)

	1970	1971	1972	1973	1974
Monetary Concepts	7,941	9,714	11,423	14,302	18,314
Current Account Deposits	4,014	4,957	5,580	7,150	9,874
Time And SavingsDeposits	4,424	5,259	6,456	7,879	9,398
Bills	- 497	- 502	- 613	- 727	- 958
External Sector	- 142	- 63	47	- 346	- 175
Net International Reserves					
Loans in Foreign Money	96	75	151	204	218
Deposits in Foreign Money	- 9	-12	- 95	- 81	- 15
Other Foreign Money Obligations	- 258	- 209	- 350	- 517	- 460
Government Sector	- 80	- 91	14	- 65	- 748
Silver, Nickel, and cupro-nickel	44	50	52	57	67
Treasury letters	276	—	—	—	—
Other Investments	23	279	355	415	594
Deposits	- 390	- 416	- 390	- 534	-1,450
Other Accounts Net	- 33	- 4	- 3	-3	41
Private Sector	7,308	8,260	9,754	12,552	16,923
Loans and Investments					
Dividends and Profits Due	- 26	- 23	- 24	- 20	- 17
Paid Capital and Reserves	-1,657	-1,826	-1,950	-2,183	-2,554
Other Accounts Net	43	36	- 10	- 166	- 365
Banking Sector	855	1,608	1,608	2,161	2,314
Central Bank Deposits	1,183	1,888	2,130	3,008	3,807
Central Bank Obligations	- 29	- 22	- 18	- 1	- 134
Commercial Banks—Debtor Banks	156	205	273	348	603
Commercial Banks—Circulating Checks	- 237	- 260	- 377	- 405	- 637
Commercial Banks—Creditor Banks	- 203	- 284	- 314	- 431	- 772
Official Bank Bonds and Mortgages	49	87	107	60	42
Official Bank Deposits	- 97	- 132	- 269	- 412	- 633
Mortgage Bank Stock	36	46	46	50	60
Mortgage Bank Mortgages	58	70	113	119	164
Mortgage Bank Deposits	- 23	- 42	- 49	- 48	- 74

Source: *Informe Económico 1974,* Banco Central de Venezuela.

Table 22. Assets and Liabilities of the Central Bank of Venezuela, 1950–1974 (million of bolívares)

	Net Int'l. Reserves	Loans & Investment	Nat'l Money and Other Assets	Total Assets and Liabilities	Bills Emitted	Deposits and Other Assets	Paid Capital and Reserves
1950	995	60	39	1095	798	285	12
1951	1110	67	45	1222	871	338	13
1952	1327	51	41	1420	1000	407	13
1953	1463	60	48	1570	1041	511	18
1954	1454	132	44	1628	1121	485	22
1955	1607	63	71	1742	1154	564	24
1956	2792	118	84	2994	1290	1679	25
1957	4246	87	113	4445	1591	2799	55
1958	3014	27	124	3165	1717	1346	102
1959	2151	253	77	2480	1709	665	106
1960	1858	471	120	2450	1758	559	133
1961	1773	617	251	2640	1632	858	150
1962	1779	778	150	2707	1581	949	177
1963	2277	731	117	3125	1653	1263	209
1964	3045	438	107	3591	1669	1642	280
1965	3060	426	122	3608	1777	1489	342
1966	2772	561	256	3589	1815	1400	374
1967	3204	429	355	3988	1990	1657	341
1968	3430	419	378	4227	2168	1677	382
1969	3486	519	524	4529	2275	1820	434
1970	3871	486	492	4849	2398	1943	508
1971	5803	136	503	6442	2574	3303	565
1972	6633	83	553	7269	2920	3733	616
1973	10080	197	572	10849	3302	5892	1655
1074	27168	455	567	28190	4391	21883	1916

Source: *La Economía Venezolana en los Ultimos Treinta Años,* Banco Central de Venezuela, Caracas, 1971, for 1950–1969; *Revista del Banco Central de Venezuela,* October–December 1974, for 1970–1974.

Table 23. Venezuelan Price Indices, 1950–1969

	Wholesale Price Index	Wholesale Prices Indices*	National Wholesale Price Index*	Imported Wholesale Price Index	Cost of Living Index	Agriculture Price Index Wholesale	Manufacturing Wholesale Price Index	Consumer Goods Wholesale Price Index	Capital Goods Wholesale Price Index
1950	100.3	164.7	163.3	155.2	90				
1951	104.6	173.6	172.4	174.7	97				
1952	103.3	176.4	172.6	171.2	98				
1953	100.0	170.9	169.5	166.6	97				
1954	101.8	175.4	172.7	162.9	97				
1955	103.0	176.5	171.0	163.7	96				
1956	102.6	76.9	85.8	64.4	98	81.7	71.9	82.5	61.4
1957	100.0	77.2	84.9	65.5	95	78.7	72.8	81.5	63.0
1958	101.8	78.3	87.3	65.6	100	85.9	72.7	83.8	62.7
1959	102.4	80.4	90.4	66.7	105	90.0	74.2	86.7	63.8
1960	101.1	81.0	90.1	68.9	109	90.1	74.7	86.7	64.5
1961	101.8	82.2	90.5	69.9	106	90.8	76.3	88.4	65.5
1962	101.7	86.2	90.3	77.4	105	89.0	81.7	89.9	73.2
1963	103.1	88.9	90.2	82.7	106	88.4	85.3	90.9	78.9
1964	106.9	92.7	95.0	85.4	107	98.1	87.9	96.7	81.4
1965	109.9	95.7	98.2	91.5	109	97.3	94.0	97.9	89.2
1966	112.0	97.0	98.1	95.4	112	06.6	97.1	98.0	93.6
1967	110.1	98.3	98.5	98.1	112	95.6	98.6	98.4	97.7
1968	113.0	100.0	100.0	100.0	114	100.0	100.0	100.0	100.0
1969	115.1	101.5	101.2	102.1	116	101.5	100.9	100.9	102.3

*The index in the second, third and fourth columns are each two indices, one for 1950–1955 for the Federal District on a 1938 base, the others for the Republic on a 1968 base. The general index in the second column was calculated independently of the other two.

Source: The index in the first column is from *Anuario Estadístico 1970,* Dirección General de Estadística, Caracas, 1972, and the others from *La Economía Venezolana en los Ultimos Treinta Años,* Banco Central de Venezuela, Caracas, 1971.

Table 24. Venezuelan Wholesale Price Index, 1970–1974
(1968 = 100)

	1970	1971	1972	1973	1974
General Index	103.1	106.7	110.4	117.4	136.8
Agricultural Products	99.9	100.5	104.9	118.5	135.4
Prepared Foods	103.8	106.4	109.9	116.4	133.4
Textiles	100.4	102.3	103.6	108.2	128.5
Hides and Skins	104.2	99.1	103.2	130.6	185.5
Energy	100.2	101.5	100.5	100.5	100.6
Chemical Products	101.0	1p1.8	103.0	105.5	115.2
Wood	111.4	131.5	133.4	140.2	151.9
Paper and Cardboard	103.6	104.3	105.5	109.8	143.1
Furniture	102.3	104.8	108.3	109.8	122.7
Machinery and Equipment	107.3	113.8	120.4	124.6	139.0
Motor Vehicles	105.8	109.2	113.5	121.1	129.2
Construction Material	99.7	102.0	104.1	110.8	147.4
Tobacco and Beverages	100.4	107.1	109.3	112.6	125.7
Electrical Equipment	102.0	104.9	104.0	132.3	192.1
Other	104.7	107.6	113.4	118.0	142.7

Source: *Informe Economico 1974,* Banco Central de Venezuela, A-115–152

Table 25. Venezuelan Imports and Trade and Service Balance, 1950–1959 (000,000$)

Merchandise Imports	Financial Services	Non-financial Services	Merchandise and Service Imports	Merchandise Balance +	Service Balance –	Current Account Balance
533	392	141	1,066	622	533	89
683	424	161	1,268	687	592	95
770	447	180	1,397	676	62u	49
858	439	236	1,533	640	675	–35
837	486	259	1,582	811	745	66
981	585	321	1,887	910	906	4
1,180	744	411	2,335	1,031	1,155	–124
1,776	977	582	3,335	975	1,558	–583
1,514	600	514	2,628	958	1,114	–156
1,527	512	463	2,502	790	976	–186
1,151	508	392	2,051	1,207	900	307
1,055	556	311	1,922	1,326	867	459
1,080	623	304	2,007	1,426	927	499
969	610	291	1,870	1,530	901	629
1,204	666	388	2,258	1,299	1,054	245
1,355	704	389	2,488	1n127	1,093	34
1,316	673	394	2,383	1,082	1,067	15
1,366	652	363	2,381	1,166	1,015	151
1,585	705	475	2,765	952	1,180	–228
1,630	689	464	2,783	893	1,153	–260

Source: *La Economía Venezolana en los Ultimos Treinta Anos,* Banco Central de Venezuela, Caracas, 1971.

Table 26. Venezuelan Sector Trade and Balances, 1950–1965 (000,000$)

	Petroleum Exports –	Petroleum Imports –	Petroleum Services –	Petroleum Balance	Iron Ore Exports +	Iron Ore Imports –	Iron Ore Services –	Iron Ore Balance	Rest of Economy Exports +	Rest of Economy Imports –	Rest of Economy Services –	Rest of Economy Balance
1950	1,124	66	425	633	—	—	—	—	31	467	108	–544
1951	1,297	119	455	723	—	—	—	—	42	564	137	–659
1952	1,384	179	477	728	—	—	—	—	62	588	150	–676
1953	1,428	144	479	805	18	32	5	–19	52	679	191	–818
1954	1,584	122	536	926	36	14	5	17	48	699	204	–855
1955	1,781	135	669	977	50	7	15	28	50	837	222	1,009
1956	2,086	237	859	990	76	12	25	39	47	931	271	1,155
1957	2,570	396	1,103	1,071	118	29	43	46	62	1,350	413	–1,701
1958	2,299	231	692	1,376	120	14	39	67	53	1,264	382	–1,593
1959	2,128	158	541	1,429	129	7	42	80	60	1,356	392	–1,688
1960	2,149	80	456	1,613	158	10	50	98	51	1,056	394	–1,399
1961	2,213	51	523	1,639	125	5	38	82	43	989	306	–1,252
1962	2,343	85	609	1,649	113	6	25	82	50	983	293	–1,226
1963	2,336	75	585	1,676	92	10	20	62	71	878	296	–1,103
1964	2,341	61	665	1,615	107	6	36	65	55	1,128	353	–1,426
1965	2,305	65	658	1,582	123	5	44	74	54	1,280	391	–1,617
1966	2,215	48	618	1,549	123	4	43	76	60	1,256	406	–1,602
1967	2,333	41	575	1,717	118	6	38	74	81	1,309	402	–1,630
1968	2,356	58	696	1,602	103	8	31	64	78	1,511	453	–1,886
1969	2,306	99	651	1,556	130	29	53	48	87	1,494	449	–1,856

Source: *La Economía Venezolana en los Ultimos Treinta Años,* Banco Central de Venezuela, Caracas, 1971.

Table 27. Venezuelan Capital Account and Balance of Payments, 1950–1969 (000,000$)

	Short-term Private Capital*	Short-term Public Capital	Total Short-term Capital	Long-term Petroleum	Long-term Iron Ore	Long-term Rest of Economy	Total Long-term Capital	Capital Account	Current Account	Capital and Current Account	Errors and Omissions	Monetary Movement
1950	–8	0	–8	–46	0	63	17	9	89	98	–178	80
1951	17	–4	13	–94	0	–102	–196	–183	95	–88	93	–5
1952	–4	4	0	71	0	84	155	155	49	204	–134	–70
1953	–1	0	–1	50	57	33	140	139	–35	104	–58	–46
1954	48	–5	43	–21	14	19	12	55	66	121	–122	1
1955	45	0	45	–20	–8	26	–2	43	4	47	0	–47
1956	24	–1	22	385	–1	79	462	486	–124	362	60	–421
1957	4	–5	–1	803	8	172	983	982	–583	399	102	–501
1958	–68	–336	–404	59	3	19	82	–322	–155	–477	92	386
1959	12	–283	–271	109	–14	–41	53	–218	–186	–404	37	367
1960	–22	–6	–28	–145	–35	108	–72	–100	307	207	–315	108
1961	–254	0	–254	–126	12	–178	–292	–546	459	–87	68	19
1962	–221	–2	–223	–304	–14	69	–249	–473	499	26	–28	2
1963	–185	4	–181	–106	1	28	–77	–258	629	371	–215	–156
1964	26	3	29	–98	–19	49	–68	–39	245	206	–110	–96
1965	–1	5	4	–93	–17	142	32	36	34	70	–83	13
1966	–65	4	–61	–160	–2	158	–4	–65	15	–50	–1	51
1967	–4	5	1	–167	–1	119	–49	–48	151	103	–1	–102
1968	4	5	9	86	5	131	222	231	–228	3	51	–54
1969	169	5	174	26	17	153	196	370	–260	110	–101	–9

Source: *La Economía Venezolana en los Ultimos Treinta Años,* Banco Central de Venezuela, Caracus, 1971, pp. 295–302.

*Does not include Petroleum and Mining Sectors

Table 28. Venezuelan Balance of Payments, 1970–1974
(millions of U.S. dollars)

	1970	1971	1972	1973	1974
Current Transactions	–56	125	–39	880	6,163
Exports	2,756	3,372	3,782	5,628	15,206
Imports	–1,713	–1,896	–2,222	–2,626	–3,841
Commercial Balance	1,043	1,476	1,560	3,002	11,365
Transport and Insurance	–164	–208	–242	–257	–352
Other Transport	–15	–20	–19	–6	0
Other Insurance	–9	–14	–14	–14	–14
Travellers	–90	–77	–105	–75	–110
Income on Investment	–669	–874	–1,062	–1,584	–4,329
Government Expenditures, nec	–14	–34	–15	–11	–20
Other Services	–46	–41	–47	–64	–183
Service Balance	–1,007	–1,268	–1,504	–2,011	–5,008
Good and Service Balance	36	208	56	991	6,357
Unilateral Transfers	–92	–83	–95	–111	–194
Non-Monetary Capital	78	508	–176	173	–693
Private	–54	323	–297	70	–496
Long-term	–37	195	–379	74	–202
Investment	–23	211	–376	–60	–191
Loans Received	–3	–7	–4	–4	–1
Stock Transactions	–11	–9	–1	–10	–10
Short-Term	–17	128	82	144	–294
Government	132	185	121	102	–197
Long-term	127	180	124	80	–179
Short-term	5	5	–3	22	–18
Errors and Omissions	21	–212	448	–510	–1,444
Balance on Capital and Current Account	43	421	233	542	4,046
Creation of Reserves	41	35	35	99	2
Monetary Movement	–84	–456	–268	–641	–4,028
Foreign Exchange	–41	–416	–167	–609	–3,974

Source: *Informe Económico,* Banco Central de Venezuela, p. A-236.

Table 29. Venezuelan Prices and Terms of Trade, 1950–1969

	Export Prices	Import Prices	Net Terms of Trade	Quantum of Exports	Quantum of Imports	Gross Terms of Trade	Capacity to Import
1950	91.4	61.4	148.9	52.4	61.8	84.8	78.0
1951	91.7	67.2	136.5	59.3	67.3	88.1	80.9
1952	95.9	70.1	136.8	63.9	72.9	87.7	87.4
1953	103.1	70.4	146.4	62.7	74.2	84.5	91.8
1954	103.7	72.8	142.4	68.0	79.0	86.1	96.8
1955	104.5	72.3	144.5	77.3	86.2	89.7	111.7
1956	105.4	77.5	136.0	88.9	94.0	94.6	120.9
1957	115.5	84.9	136.0	101.1	139.4	72.5	136.0
1958	108.4	84.8	127.8	98.1	119.3	82.2	127.8
1959	100.0	100.0	100.0	100.0	100.0	100.0	100.0
1960	96.0	102.0	94.1	105.2	74.2	141.8	99.0
1961	96.2	102.7	93.7	106.4	66.8	154.3	99.7
1962	93.4	107.7	86.7	115.2	65.5	175.9	99.9
1963	90.9	97.6	93.1	117.0	64.0	182.8	108.9
1964	87.2	93.7	93.1	123.2	84.0	146.7	114.7
1965	85.0	100.3	84.7	125.3	87.7	142.9	106.1
1966	85.3	102.6	81.4	122.7	83.1	147.7	99.9
1967	83.3	101.0	82.5	128.7	87.8	146.6	106.2
1968	83.5	100.5	83.1	128.5	102.4	125.5	106.8
1969	81.3	104.4	77.9	131.4	101.8	129.1	102.4

Source: *La Economía Venezolana en los Ultimos Treinta Años,* Banco Central de Venezuela, Caracas, 1971.

Table 30. Venezuelan Export Volume and Price and Import Price Index 1969–1975 (1970 = 100)

	1969	1970	1971	1972	1973	1974	1975
Petroleum Exports	101.8	100.0	94.5	89.4	92.4	79.3	63.6 (Sept)
Crude Petroleum Exports	101.7	100.0	95.0	87.8	87.2	72.8	68.9 (Sept)
Refined Product Exports	90.4	100.0	93.5	93.0	103.8	93.5	51.9 (Sept)
Iron Ore Exports	91.0	100.0	92.0	79.0	104.0	126.0	103.0 (Nov.)
Petroleum Export Price	100	100	100	112	126	381	501 (Sept)
Iron Ore Export Price	98	100	106	112	107	134	192 (June)
Import Price Index	97.1	100.0	105.3	109.7	115.6	134.4	153.3 (Oct.)

Source: *International Financial Statistics,* March, 1976, pp. 412–413.

Table 31. Venezuelan International Reserves, 1950–1975
(000,000$)

	Gold	IMF Position	Net Deposits and other Assets	Total Central Bank Holdings	Commerical Bank Holdings	Total Reserves	Gold Reserve Ratio	Reserve/Import Ratio
1950	340	0	–15	326	16	342	99.4	32.1
1951	373	0	–10	363	16	379	98.4	29.9
1952	373	0	60	433	14	448	83.3	32.1
1953	373	0	104	477	16	494	75.5	32.2
1954	404	0	71	475	18	492	82.1	31.1
1955	404	0	121	525	15	539	75.0	28.6
1956	572	0	337	909	18	927	61.7	39.7
1957	654	0	727	1381	15	1,396	46.8	41.9
1958	654	0	329	983	28	1,011	64.7	38.5
1959	655	0	48	703	6	709	92.4	28.3
1960	401	37	159	597	7	605	66.3	29.5
1961	401	37	133	571	14	585	68.5	30.4
1962	401	37	138	576	7	583	68.8	29.0
1963	401	37	302	740	0	740	54.2	39.6
1964	401	37	387	825	10	835	48.0	37.0
1965	401	63	371	835	18	853	47.0	34.3
1966	401	63	304	768	4	772	51.9	32.4
1967	401	73	393	867	7	874	45.9	36.7
1968	403	81	433	917	11	928	43.4	33.6
1969	403	89	438	930	9	939	42.9	33.7
1970	384	117	520	1,013	8	1,021	37.6	59.6
1971	425	121	976	1,502	20	1,522	27,9	80.3
1972	425	121	1,186	1,653	79	1,732	24.5	77.9
1973	472	134	1,816	2,401	11	2,412	19.6	91.9
1974	479	491	—	6,494	19	6,513	7.4	174.1
1975	458	942	7,461	8,861	—	8,861	5.2	—

Source: *La Economía Venezolana en los Ultimos Treinta Años,* Banco Central de Venezuela, Caracas, 1971, pp. 243–249, 1950–1969; *International Financial Statistics,* March 1976, pp. 410–413; *Informe Económico 1974,* Banco Central de Venezuela, pp. A-244, 1970–1975

Table 32. Venezuelan Production of Selected Products, 1950–1974

	Corn (000MT)	Rice (000MT)	Yucca (000MT)	Potatoes (000MT)	Sesame (000MT)	Cotton (000MT)	Coffee (000MT)	Cacao (000MT)	Bananas (000 units)	Sugar (000MT)	Tobacco
1950	310.3	38.6	137.6	29.4	5.3	8.3	34.0	16.9	448.1	654.8	7.
1951	312.9	40.0	188.5	32.0	2.2	13.3	43.3	17.0	524.6	588.4	10.
1952	343.2	49.3	209.1	23.8	1.6	14.2	54.0	16.0	553.0	843.8	10.
1953	334.9	57.9	253.2	33.6	7.0	12.8	44.8	16.0	540.1	933.6	8.
1954	326.3	102.3	193.5	40.4	7.0	16.0	46.9	18.4	569.0	1217.7	5.
1955	317.4	60.0	156.8	45.0	9.1	13.0	53.0	17.2	690.7	1872.6	7.
1956	350.1	47.0	182.9	70.0	13.0	13.7	46.1	20.5	703.1	3056.4	8.
1957	340.1	21.8	190.3	106.0	12.7	14.2	51.0	17.5	808.0	3246.2	5.
1958	357.6	19.1	189.1	69.8	21.0	21.9	60.1	19.3	656.3	1557.3	5.
1959	336.4	38.6	217.9	93.1	19.8	24.6	55.3	17.5	800.5	1652.5	6.
1960	439.5	71.9	340.2	133.6	16.2	29.6	59.0	18.5	889.1	1924.5	9.
1961	419.5	80.7	139.1	74.0	24.9	36.4	57.1	16.9	860.5	3000.7	10.
1962	540.5	103.1	322.8	121.2	28.1	32.2	54.2	18.9	788.5	3122.6	8.
1963	430.2	131.1	344.2	110.9	30.9	29.4	60.7	20.8	792.4	3299.7	8.
1964	475.0	165.8	311.7	123.5	46.6	35.7	56.1	20.9	821.1	3705.9	8.
1965	521.0	199.9	301.4	135.9	54.1	41.5	54.4	21.9	812.1	4213.3	9.
1966	557.5	195.0	320.0	126.1	60.0	35.3	61.0	22.9	798.7	4428.1	9.
1967	633.4	223.1	327.7	132.6	80.5	37.4	61.3	23.1	742.2	4585.4	9.
1968	660.8	244.6	340.9	142.5	76.2	36.7	59.2	24.0	836.6	4051.7	11.
1969	670.3	243.9	309.8	123.8	82.6	41.1	60.7	18.3	980.2	4216.8	9.
1970	709.9	226.2	317.2	125.0	125.6	40.0	60.6	18.9	968.5	4908.7	11.
1971	713.5	153.0	322.7	115.2	93.9	42.6	58.4	19.6	988.9	5152.1	13.
1972	506.3	164.7	318.2	109.4	59.2	57.3	40.4	16.7	997.0	5475.5	12.
1973	454.4	302.0	272.1	123.6	60.0	60.0	65.9	19.1	902.2	5623.1	13.
1974	683.6	292.6	317.8	149.8	76.2	76.2	44.2	17.4	952.3	6069.5	14.

Sources: *Anuario Estadístico, 1969,* p. 157; *1966,* p. 157, *1965,* 163, *1955–1956,* 201; *La Economi Venezolana en las Ultimos Treinta Años,* Banco Central de Venezuela, Caracas, 1971, p. 104, 113; *Inform Económico 1974,* Banco Central de Venezuela, p. A-135—A-142

Table 32. Venezuela Production of Selected Products, 1950–1974 (*continued*)

	Milk (000,000 liters)	Cattle (000 heads)	Fowl (000,000 heads)	Eggs (000,000 units)	Meat (MT)
1950	177.0	449.0	2.0	62.0	71.0
1951	196.0	497.0	2.0	64.0	71.0
1952	231.0	517.0	4.0	69.0	71.0
1953	261.0	545.0	5.0	73.0	76.0
1954	289.0	540.0	7.0	79.0	78.0
1955	316.0	558.0	7.0	79.0	82.0
1956	318.0	576.0	8.0	81.0	85.0
1957	320.0	634.0	9.0	86.0	96.0
1958	373.0	721.0	11.0	92.0	110.0
1959	375.0	754.0	16.0	165.0	116.0
1960	421.0	794.0	20.0	423.0	121.0
1961	444.0	864.0	31.0	606.0	137.0
1962	481.0	879.0	34.0	756.0	138.0
1963	521.0	929.0	37.0	669.0	147.0
1964	586.0	1,004.0	45.0	837.0	158.0
1965	626.0	1,030.0	46.0	930.0	162.0
1966	661.0	1,115.0	48.0	1,33.0	180.0
1967	708.0	1,149.0	54.0	1,065.0	183.0
1968	730.0	1,181.0	75.0	1,155.0	184.0
1969	767.0	1,230.0	61.0	1,321.0	208.4
1970	923.5	1,116.5	65.0	1,303.9	200.9
1971	969.9	1,102.2	68.4	1,371.7	214.1
1972	1,048.6	1,228.4	77.8	1,362.7	216.9
1973	1,023.2	1,183.3	86.4	1,547.5	230.5
1974	1,282.4	1,282.4	99.4	1,431.2	239.6

Table 33. World Production of Crude, 1950–1974
(000 barrels per day)

	1950	1955	1960	1965	1969	1970	1971	1972	1973	1974
United States	5,407	6,807	7,035	7,753	9,210	9,630	9,530	10,150	9,190	8,800
Canada	80	355	524	788	1,126	1,260	1,350	1,540	1,800	1,690
Venezuela	1,498	2,157	2,846	3,473	3,594	3,710	3,550	3,220	3,370	2,980
Kuwait	344	1,091	1,624	2,170	2,576	2,740	2,890	3,000	3,020	2,550
Iran	664	330	1,068	1,859	3,375	3,830	4,540	5,020	5,860	10,030
Saudi Arabia	547	965	1,247	1,996	2,993	3,550	4,490	5,730	7,610	8,480
Iraq	136	688	967	1,278	1,526	1,570	1,700	1,450	1,950	1,850
USSR	729	1,397	2,952	4,752	6,560	7,620	8,470	8,910	10,020	10,780
Libya	—	—	—	1,239	3,111	3,320	2,760	2,220	2,190	1,490
Algeria	—	1	184	547	943	1,010	780	1,060	1090	990
Others	1,034	1,622	2,567	4,135	6,517	7,210	8,210	8,440	9,730	10,210
Totals	10,419	15,413	20,994	29,990	41,531	45,450	48,270	50,740	55,830	55,850

Source: *La Economía Venezolana en los Ultimos Treinta Años,* Banco Central de Venezuela, Caracas, 1971, 1950–69; *Informe Económico, 1974,* Banco Central de Venezuela, p. A-168, 1970–1974.

Table 34. World Reserves of Crude 1950–1974
(000,000,000 barrels)

	1950	1955	1960	1965	1969	1974
United States	26	30	32	31	39	35
Venezuela	9	12	17	17	15	19
Middle East	48	133	163	288	317	404
Soviet Bloc	6	10	25	37	60	83
Africa	0	0	8	23	55	56
Rest of the World	6	12	21	14	38	122
Total	95	197	266	350	524	719

Sources: *La Economía Venezolana en los Ultimos Treinta Años,* Banco Central de Venezuela, Caracas, 1971, 1950–1969; *Informe Económico,* Banco Central de Venezuela, A-167 for 1974.

Table 35. Venezuelan Crude, Refined Products and Natural Gas Production 1950–1974

	Crude Production 000 b/d	Rate of Growth	Refined Petroleum 000 b/d	Refined Crude (%)	Natural gas (000,000,000M³)
1950	1,498		250	16.7	16
1951	1,705	13.8	315	18.5	19
1952	1,804	5.8	347	19.2	21
1953	1,765	–2.2	413	23.4	21
1954	1,895	7.4	443	23.4	22
1955	2,157	13.8	537	24.9	24
1956	2,457	13.9	634	25.4	28
1957	2,779	13.1	689	24.8	32
1958	2,605	–6.3	732	28.1	32
1959	2,771	6.4	824	29.7	32
1960	2,846	2.7	882	31.0	32
1961	2,920	2.6	928	31.8	33
1962	3,200	9.6	1,025	32.0	36
1963	3,248	1.5	1,042	32.1	37
1964	3,393	4.5	1,092	32.2	39
1965	3,473	2.4	1,175	33.8	41
1966	3,371	–2.9	1,174	34.8	41
1967	3,542	5.1	1,166	32.9	46
1968	3,605	1.8	1,186	32.9	46
1969	3,594	–0.7	1,159	32.2	47
1970	3,708	3.2	1,282	34.6	48
1971	3,549	–4.3	1,198	33.8	48
1972	3,220	–9.3	1,192	37.0	46
1973	3,366	4.5	1,330	39.5	49
1974	2,976	–11.6	1,198	40.3	46

Sources: *La Economía Venezolana en los Ultimos Treinta Años,* Banco Central de Venezuela, Caracas, 1971, 1950–1969; *Informe Económico 1974,* Banco Central de Venezuela, A-160, 1970–1974.

Table 36. Venezuelan Petroleum Financial Indicators, 1947–1974
(millions of bolívares)

	Total Income	Income Tax	Royalties	Total to Government	Profits After Taxes	Government Share of Total Income
1947	2,394	297	397	818	745	52
1948	3,534	479	640	1,290	1,060	55
1949	3,124	272	627	1,055	704	60
1950	3,748	394	519	1,021	970	51
1951	4,405	525	727	1,448	1,201	55
1952	4,677	594	751	1,544	1,262	55
1953	4,892	507	786	1,502	1,261	54
1954	5,337	585	874	1,576	1,412	53
1955	5,875	712	1,003	1,841	1,710	52
1956	6,829	931	1,188	2,281	2,115	52
1957	8,463	1,199	1,550	2,990	2,774	52
1958	7,662	1,465	1,415	3,067	1,616	65
1959	7,284	1,260	1,444	2,860	1,335	68
1960	7,287	1,070	1,503	2,711	1,282	68
1961	7,477	1,216	1,552	2,899	1,477	66
1962	7,703	1,462	1,703	3,225	1,694	66
1963	7,701	1,544	1,731	3,331	1,679	66
1964	10,693	2,251	2,557	4,862	2,457	66
1965	10,725	2,323	2,564	4,937	2,638	65
1966	10,419	2,260	2,531	4,836	2,504	66
1967	10,964	2,752	2,663	5,460	2,514	68
1968	11,119	2,754	2,715	5,513	2,653	68
1969	10,906	2,751	2,722	5,526	2,264	71
1970	11,384	3,270	2,875	6,20	1,739	78
1971	13,720	4,653	2,836	7,546	2,247	77
1972	13,566	5,558	2,797	8,411	1,266	87
1973	19,178	8,828	3,496	12,410	2,812	82
1974	45,354	28,730				

Table 36. Venezuelan Petroleum Financial Indicators, 1947–1974 (millions of bolívares) (*continued*)

Net Profit Share of Total Income	Gross Accumulated Investment	Depreciation & Amortization	New Investment	Net Accumulated Investment	Net Profits Net Accumulated Investment
31	5,320	190	1,127	3,305	22.5
30	6,697	246	1,630	4,422	24.0
23	7,641	325	1,127	5,033	14.0
26	8,027	420	561	5,021	19.3
27	8,526	495	727	5,142	23.4
27	9,312	530	967	5,457	23.1
26	9,873	558	901	5,460	23.1
26	10,212	613	933	5,534	25.5
29	10,937	680	928	5,649	30.3
31	13,038	732	1,232	7,023	30.1
33	15,828	812	1,822	9,002	30.8
21	17,218	836	1,788	9,652	16.7
18	18,803	939	1,262	10,375	12.9
18	19,987	948	730	9,771	13.1
20	19,155	929	516	9,140	16.2
22	19,417	1,082	474	8,664	19.6
22	19,353	986	511	8,086	20.8
23	19,872	993	753	7,778	31.6
25	20,342	965	825	7,553	34.9
24	20,518	991	638	7,043	35.6
23	20,673	974	647	6,576	38.2
24	21,635	938	1,182	6,853	38.7
21	22,655	1,009	1,574	7,361	30.8
16	23,647	1,070	1,294	7,461	23.3
16	23,732	1,211	1,287	7,434	30.2
9	24,735	1,140	1,000	7,334	17.3
15	25,565	1,079	864	7,122	39.5
	26,273	1,027			

Note: Investment data do not include concession payments. Investment data prior to 1964 is not revalued at depreciated exchange rate of that year, thereby inflating the profit/investment ratio.

Source: *Petróleo y Otros Datos Estadísticos,* various issues Ministerio de Minas e Hidrocarburos and *Informe Económico,* various, Banco Central de Venezuela.

INDEX